SPREADSHEET PHYSICS

Charles W. Misner
University of Maryland, College Park

Patrick J. Cooney
Millersville University, Pennsylvania

Addison-Wesley Publishing Company

Reading, Massachusetts • Redwood City, California • New York
Don Mills, Ontario • Wokingham, England • Amsterdam • Bonn • Sydney
Singapore • Tokyo • Madrid • San Juan • Milan • Paris

To Margaret, Mary, and Jay   *PJC*

To Susanne   *CWM*

Contents

List of Figures

List of Screens

Preface

The rapid spread of electronic spreadsheets through business and commercial offices during the past decade is a Darwinian demonstration of the spreadsheet's fitness for many common tasks. It should not be too surprising then that these demonstrated merits have applications in science also. Thus many a scientist who originally bought a spreadsheet for income tax calculations at home later found uses for it in the office and laboratory. The appropriate uses in science tend to be quick looks at relatively simple or idealized siutations involving limited data—just the class of scientific problems that are chosen as examples for teaching purposes.

Our original aim was to provide written materials that a student in an introductory university physics course could be given, allowing a modified course that used some numerical methods. That aim remained the central one, but the same material can serve other ends, such as introducing spreadsheets to technically oriented students. Our enthusiasm for the improved physics that becomes accessible with these techniques, however, has led to the inclusion of many additional topics. These more advanced topics appear in optional sections and are suitable either for student projects in introductory courses or as numerical experiments to complement advanced analytical techniques in intermediate (junior/senior) mechanics courses.

As a result, this book can be used in several different ways by readers with different interests. Our intended audience includes students in introductory physics courses, students at more advanced levels who want to learn to use a spreadsheet in a scientific context, teachers and student teachers who want to prepare for using spreadsheets in high school and middle school science, and physics students in intermediate mechanics courses.

Learning to Use a Spreadsheet

For the technically oriented student the first few chapters provide a quick introduction to the use of a spreadsheet. An advanced student who already is familiar with the MS-DOS operating system, a word processor or editor, and a couple of programming languages can work through the first two chapters of this book in an afternoon and acquire a useful, but not skillful, command of the spreadsheet. These two chapters already provide the ability to tabulate and graph functions and to solve differential equations and graph the results. The student in an introductory physics course faces much more unfamiliar physics and computer skills in these two chapters and may spread the work over some weeks. Additional material in the other chapters of Part I provides improved skills in basic spreadsheet use and more efficient numerical methods, as well as the foundations of Newtonian mechanics. A few additional skills are practiced in the applications in Part II; the student who wishes primarily to learn spreadsheet skills in a technical context should look under "spreadsheet skills" in the index for pointers to applications and exercises where these additional spreadsheet topics are treated.

Introductory University Physics

This volume is intended to be used in the Introductory University Physics course (calculus-based introductory physics for scientists and engineers) in parallel with standard texts such as SEARS, ZEMANSKY, AND YOUNG, *University Physics* (Addison-Wesley) or HALLIDAY AND RESNICK, *Fundamentals of Physics* (Wiley). Its goal is to multiply the topics that can be discussed by providing a powerful, easy-to-use, mathematical tool that allows solving nonlinear differential equations in the introductory course.

New topics, of course, cannot be introduced without omitting some traditional material, and new skills cannot be taught without providing time for experimenting with those skills. In some courses the first laboratory hours provide an appropriate setting for working on Chapters 1 and 2. In all courses the parts of this book that are used must replace either similar subject matter from a traditional course or topics that are less important for the begining student. The major trade we suggest is to drop rigid body theory such as moments of inertia and related static equilibria while deepening the study of point particle motions. An additional possibility is to omit the dynamics of fluids (e.g., the Bernoulli equation).

This book does not try to be a complete course in either text or scope and assumes a conventional text as backbone for the course. We treat the computer as a powerful, problem-solving tool that students will learn to use on a small scale in the same way that scientific researchers and practicing engineers use it in their work—to solve problems and construct mathematical models that could never be approached without the computer. Students are taught to make direct "sledge-hammer" attacks on problems that, in the past, would have required years of sophisticated mathematical studies to crack. The basic laws of physics are used in their pure, raw form instead of in forms diluted and confused by specialization to particular problems.

Pre-college Teachers

Widely felt dissatisfaction with the intellectual skills of much of the United States workforce seems likely to lead to significant changes in educational practices. Teachers at the pre-college level, and students intending to teach there, may wish to prepare for such changes or to participate in their design. We believe that the use of spreadsheets by students at the high school and middle school level could make science and mathematics there more interesting, more accessible, and more insightful. This belief is based on the spreadsheet's ability to describe and model quantitative change without burdensome prerequisites of abstract mathematics.

The description and modeling of quantitative change together comprise an important tool for many sciences, including not only physics, chemistry, biology, geology and meteorology, but also social sciences such as economics. This tool is currently taught primarily in courses on differential equations and their applications. Prerequisites for such courses include algebra and calculus. But spreadsheets are widely used to describe and model change by workers who do not consider themselves skilled in algebra and calculus. The spreadsheet enables this through its ability to handle numbers in wholesale lots without demanding a knowledge of the abstract concepts of arrays or functions and through its ability to display easily relationships among such collections of numbers by drawing graphs.

While the presentation in this book presumes the increasing algebra and calculus skills demanded of a first-year engineering or science student, the computations in most cases can be understood at a more elementary level. For example, in Newton's law—force causes a change in velocity—the computations can be regarded as just simple multiplication and addition repeated many times. In a similar spirit, we hope that imaginative teachers will translate some of the best ideas in science into arithmetic, words, and ideas for younger students. Then, with spreadsheets to handle the repetitious arithmetic, students would see good science examples first and only later, motivated by these examples, find that abstract mathematics is an important tool for communicating the ideas they form by exploring such examples.

Notations and Conventions

Exercises in this book are classified in four categories:

Core[C] An exercise that reinforces the main points being taught in that chapter. These exercises should be done by anyone seriously studying the section in which they appear. Usually they simply say "Do what you were told to do in the section you just read." The point of this is to allow the instructor to assign these exercises as homework—they also specify the kind of material that the student is expected to turn in for marking.

Parallel[P] An exercise that, with variations, covers the same type of material as the core exercises. These exercises can be assigned by an instructor for student homework in instances where the corresponding core exercise has been worked out in lecture demonstration.

Extension[E] An exercise that goes beyond the basics and teaches material that is not particularly more difficult than the basics but requires additional time to learn and practice.

Honors[H] An exercise that is substantially more challenging than those in other categories. Some honors exercises may require more analytical mathematics skills than core exercises. Others will simply be large problems that are more suitable for a several-week project than for a simple homework assignment.

The category of each exercise is shown by a superscript following the exercise number, as in EXERCISE 0.0^{C}.

A convention that we try to use consistently is to let the word **spreadsheet** refer to a large, professionally written computer program such as Lotus *1-2-3*, Borland *Quattro*, or Microsoft *Excel* that gives a computer the personality you get acquainted with in this book. The word **worksheet** refers to a file that the student or other user of a spreadsheet produces in the course of solving a problem; it contains the formulae specific to that problem, the input and output data, and the specifications for the graphs that present some of that data.

Typography

Several different type styles are used that have special significance.

`Typewriter Type` is used for characters that you are to enter from your computer keyboard, or which you may read on your computer screen.

{Keyname} is the way we refer to any special key on your computer keyboard that does not type a standard character. The most important of these are the {Enter} key that tells the computer to act on what you have just typed and the {F1} key by which spreadsheets provide helpful information on their use. To emphasize this latter function, we will often refer to the {F1} key as {Help}.

Boldface is used in three different contexts. In ordinary text it identifies a **technical term** that is being defined at that point. It is also used to denote constants in Schemata, which are worksheet outlines in mathematical notation (cf. pp. 11 and 62). Where physics rather than spreadsheets is the focus of attention, boldface mathematical symbols are used to represent vectors.

Environment

So that instructions can be clear and specific, the book is being written for one hardware/software environment—IBM personal computers and Lotus *1-2-3*. The many versions of IBM PCs (plain, XT, AT, PS/2) and their clones are all satisfactory, as are spreadsheet programs that closely emulate Lotus *1-2-3*. The use of this material in courses requires computer facilities for the students and, ideally, for demonstrations in the classroom.

♠ Optional Sections

This book is divided into two major parts: Fundamentals, and Applications. These terms refer to spreadsheet skills and numerical analysis only; much fundamental physics is introduced in the Applications group of chapters. Applications chapters are, however, optional in the sense that none is a prerequisite to another in the development of computer skills for modeling physical systems. The chapters in the Fundamentals part, in contrast, are expected to be studied in order and the main parts of them are used in most later chapters.

Some chapters need not be studied in full. The optional sections are marked with a special symbol as in the heading of this section. These sections can be omitted without loss of continuity (even in the Fundamentals chapters). Many of the optional sections discuss topics that would be suitable for student projects. Most of them would also be suitable for study and for lecture demonstrations in intermediate mechanics courses.

Acknowledgements

The approach to physics teaching illustrated by this book was encouraged initially by an equipment grant to the University of Maryland by IBM under its Advanced Educational Projects program. Subsequent support for this and related projects of the Maryland University Project on Physics and Educational Technology was received from the Fund for the Improvement of Post-Secondary Education; we benefitted greatly from interaction with colleagues in this M.U.P.P.E.T. group. We have also received help and encouragement from many colleagues we are unable to list individually here; their contributions are deeply appreciated. We give special thanks to Dale Parish who reviewed major parts of the manuscript and has smoothed and clarified the writing in a valuable way. The screen photograph for the cover is by Joan Wright Hamilton. The friendly encouragement and knowledgeable support from Stuart Johnson and his associates at Addison-Wesley have made completing the book less agonizing than anticipated.

Part I

Fundamentals

Chapter 1

The Basics of Spreadsheets

Electronic spreadsheets are powerful tools that have revolutionized problem solving and analysis in today's world. An electronic spreadsheet is a computer program that turns your computer screen into a smart piece of paper. It allows calculations to be done in the format traditionally used by accountants and computational physicists who stored their work with pen on paper, using mechanical desk calculators to do addition and multiplication. The program provides an on-screen image of the bookkeeper's columnar paper and supplies an "invisible assistant" who does all the repetitive and laborious work once you give instructions for the required calculations. The assistant will also graph the results if you point out what you want graphed.

Preview

This chapter is designed to introduce you to the basics of electronic spreadsheets. The best way to learn to use an electronic spreadsheet is to play a practice game. You do a sample computation by following instructions more or less blindly so that you have a little experience of the possibilities a spreadsheet offers. After that, you can begin to study the various capabilities systematically. The practice game we recommend is to fill out the computer worksheet you will name FIRST, which is shown as Screen 1.1 on page 4. After you have created FIRST, you will request the graph it creates, which is shown in Figure 1.2 on page 14.

1.1 FIRST: Your First Worksheet

The hardest part of completing the FIRST worksheet may be turning the computer on and getting the spreadsheet running. Roughly speaking, you must power up (turn on) the computer and its monitor (screen), load an operating system, and load your spreadsheet program. Please review and follow the start-up instructions in the user's manual provided with your spreadsheet. If

A7: @PI*A6   READY

	A	B	C	D	E	F	G	H
1	your name							
2	your class & section			FIRST, a simple worksheet.				
3	today's date			Find that you can type, calculate,				
4				and graph what you want.				
5								
6	0.5							
7	1.570796							
8								
9			t		sin t			
10			0	0				
11			0.2	0.198669				
12			0.4	0.389418				
13			0.6	0.564642				
14			0.8	0.717356				
15			1	0.841470				
16			1.2	0.932039				
17			1.4	0.985449				
18			1.6	0.999573				
19			1.8	0.973847				
20			2	0.909297				

FIRST.WK1

Screen 1.1: Your **FIRST** worksheet completed.

you are using a computer in a lab or computer workroom associated with a course that uses this book, your course instructor will provide the start-up instructions. Once the spreadsheet is running a blank worksheet should appear. This is a screen that is mostly empty except for numbered rows and alphabetically labeled columns. A few lines of text outside this blank area give additional information you will gradually learn to use.

Labels

If you had a real sheet of lined paper on your desk and were instructed to write "FIRST, a simple worksheet" near the middle of the second line, you would bring your pen or pencil to the position at which you wanted to begin writing. On the computer screen, this action is accomplished by using the four cursor keys (marked by arrows). On many keyboards, they are found at the right side on the keys resembling a calculator keypad. We shall, when necessary, refer to these cursor keys by the names {Up}, {Down}, {Left}, and {Right} for the keys (marked with arrows in the specified directions) that move the active position on the screen. Press these keys experimentally to see their effect on the worksheet screen in front of you. A highlight called the **cell pointer** should move in the direction you demand, except that it cannot move off the left or top edge of the sheet. In this case, the "can't do" bell or beep sounds. Use your control of the cell pointer to move it to row **2** and column **D**. You can judge your exact position on the worksheet not only by reference to the row and column labels along the edges, but also by the **content line** which is usually the top line of the screen (outside the worksheet area of labeled rows and columns). When you are where you want to be, this line begins with the cell address `D2: `. When you are in the upper left corner of the worksheet, it reads `A1: `.

After moving the cell pointer highlight to cell **D2** by using the cursor keys, type the text line that begins there in Screen 1.1, namely `FIRST, a simple worksheet`. As you type, the characters do not get written immediately in the cell where you want them, but instead appear on the **active line**, which is usually located above the worksheet area and just below the **content line**. Where a small cursor appears in the active line, you have a chance to correct any typing errors by using the `{Backspace}` key to erase characters. When you have correctly typed what you want, press `{Enter}` to signal that you are done. The line you have typed will then be displayed at the cell pointer. Each cell can hold up to 240 characters. If the text in the cell exceeds the displayed width of the cell, the spreadsheet will show as much as it can without covering the contents of another cell. To revise or correct it now, you just type it in again; when you press `{Enter}` again, the new typing will replace the old in the cell at the cell-pointer highlight. With the cell pointer still on cell **D2**, the content line should now read `D2: 'FIRST, a simple worksheet`. You should fill in all the other nonnumerical text shown in Screen 1.1 in a similar fashion.

The following **prescription** is a less verbose way of presenting the instructions in the preceding two paragraphs:

> **D2:** `FIRST, a simple worksheet.`
> **D3:** `Find that you can type, calculate,`
> **D4:** `and graph what you want.`
> **C9:** `t`
> **D9:** `sin t`
> Enter the literal text specified for these five cells.

> **A1:** *your name*
> **A2:** *your class and section*
> **A3:** *today's date*
> Supply the requested information in these cells. (The date could cause problems; read on.)

Numbers

Numbers can be entered in the worksheet just as easily as text. Screen 1.1 suggests that you place the number 0.5 in cell **A6**, which you do by moving the cell pointer to **A6** using the cursor keys, and then typing `0.5`, followed by a press of the `{Enter}` key. Since the computer actually stores the number internally in binary form to make it available for rapid calculation, it must guess whether you are doing mathematics or typing text when you begin to type. It does this on the basis of the first character typed into each cell. Thus, were you to enter a date in cell **A3** in the format **27 Jan 1991**, the spreadsheet program would beep at you to show its confusion. It assumed (from the **2**) that you were entering a number or mathematical formula and then obstinately tried to interpret **Jan** as an operator or function that it failed to find in its library. When the stupid computer beeps, you can give it another chance by pressing the {Escape} key (usually marked Esc) one or more times until a **READY** mode is indicated on the top or bottom line of the screen. To explain to the spreadsheet that **27 Jan 1991** is not a mathematical formula, precede it with a single or double quote mark, to specify text; thus, type **'27 Jan 1991**. Had you typed **1-27-91** for the date, the spreadsheet would happily display **-117** after doing the subtractions for you. Typing **'1-27-91** gives the result you want; the digits are interpreted as text characters for display, not numbers for computation.

Formulae

Displaying text and numbers on a spreadsheet is merely using it as a rather complicated typewriter, although it does make communicating the results of serious computations convenient. But it is the serious computations that you need to learn to do. A first, tiny calculation can be done in cell **A7**, as shown in Screen 1.1, by the following prescription:

> **A6:** 0.5
> **A7:** @pi*a6
>> In numbers and formulae, no spaces are allowed anywhere; lower-case letters are equivalent to upper-case letters.

Although a number is displayed at the **A7** location on the screen, you can see from the content line at the top of Screen 1.1 that the actual content is **@PI*A6**. The *at* symbol **@**, which is @ in another typeface, tells the spreadsheet that the next characters are the name of a built-in function it can find in its library, while the * symbol is the multiplication operator. Thus the formula **@PI*A6** means "multiply π by the number in cell **A6**." The **@PI** function is one whose value is always the constant 3.141592653589793; other built-in functions such as **@TAN** or **@LOG** require arguments, as in **@SIN(@PI/6)** which evaluates to 0.5. Multiplication is, of course, commutative, but if you type **a6*@pi** you will not calculate a number, but just display those characters—the spreadsheet has incorrectly guessed your intentions, based on a not particularly mathematical first character. To get it to calculate the product, you type **+a6*@pi** (with a leading plus sign).

Changing Input Data

Easy recalculation is one of a spreadsheet's major attractions. You can change the numbers in a calculation and get new answers without rewriting the formulae. To see this, enter a new number, such as 1 or 2.5 or 10, in cell **A6**. ("Enter the number 2.5 in cell **A6**" is the conventional brief way of saying that you should use the cursor keys to move the cell pointer to cell **A6** and then type the keystrokes **2.5{Enter}**.) You will see that each time you enter a new number in cell **A6**, the formula in cell **A7** is recalculated to show the result of using this new number. This recalculation will also happen when a worksheet contains hundreds or thousands of formulae that directly or indirectly depend on changed data, although it then takes a noticeable amount of time. We divert next to a discussion of saving your work and then turn to producing formulae by the hundreds.

Saving Your Work

This is a good point to stop for a minute and save your work. Unfortunately, doing this for the first time is a bit troublesome; if you are using this book in a course and are not prepared with a blank disk and your instructor's advice, you may have to skip this until another session at the computer and should proceed now to the next topic. This worksheet is small enough that recreating it from scratch is only a few minutes work the second time you do it, even though sorting out all the new ideas—cursor keys, the destructive {Backspace}, {Escape}ing from confused situations, starting the computer, etc.—take quite a time to work through at first.

Saving on a New Disk The normal save procedure is simply to type the keystrokes **/fs{Enter}**, which are abbreviations for **/File Save** *filename*, but it will not work the first time. Even more bothersome is probably not having a formatted disk at hand unless you have had a good bit of previous computer experience, so let us begin with that. Take a new floppy disk and put it in either drive **A:** or drive **B:**. (Either your course instructor or the manuals that came with your personal computer can show you what a floppy disk drive is, where it is located on your computer, and how to orient your disk correctly and close the drive door.) You must now temporarily adjourn the spreadsheet session without discarding your work by giving the command **/System**, which is done by first pressing {Escape} as many times as necessary to be sure the **READY** mode is displayed on the screen, then pressing **/** (the same key that when shifted gives the question mark), and finally pressing the **s** key. Your computer screen should now clear and show the symbol **C>** or **A>** or some other prompt that indicates you are in position to request services from the computer's operating system. Either your course instructor or the DOS manual that came with your personal computer will explain the details needed so that the **FORMAT** program is available and applies to your new floppy disk (and so that you don't erase the operating-system disk) and will prescribe a command such as **format b: /v** that will cause your new disk to be prepared for its first use. Follow the instructions on the screen only if they are what you expect; otherwise, get help. When you are asked to enter a label for your disk, type in something like **JONES1PHYS** to identify that disk both to you and to your instructor as your first disk for your physics class. Since only eleven characters are allowed in a disk label, you may have to abbreviate your last name in the label. You should never format this disk again unless you want to erase *everything* on it. Put a paper label with your name and course on the disk (write with felt pen) so that you can recognize this disk and not confuse it with a new blank one. After successfully formatting a disk, return to your worksheet by entering the word **exit** at the operating system prompt.

Saving a New File   You can now save the worksheet you have created in the computer's volatile memory in a magnetically fixed form on your disk by issuing the **/File Save** command. Type the keystrokes **/fsB:FIRST{Enter}** if your disk is in disk drive **B:**, or **/fsA:FIRST {Enter}** if it is in drive **A:**.

Taking a Break

Having saved your work, you might want to take a break at this point. The procedure for doing that (when the spreadsheet shows **READY**) is to give the command **/Quit Yes** by typing the keystrokes **/qy**. Then put your disk away and do such other cleaning up and powering down as local practice requires. To begin again, start the computer and, from the operating system prompt, load the spreadsheet program just as you did the first time. Then put your own disk containing your saved worksheet back in the same disk drive (**A:** or **B:**) you used to save your worksheet and give the command **/File Retrieve** *filename* by typing the keystrokes **/frA:FIRST{Enter}** or **/frB:FIRST{Enter}**. (No characters need to be upper case when giving these commands. We have used lower case for the command abbreviation **/fr**, which is always the same, and upper case for the file location **A:** or **B:** and the name **FIRST**, which changes when you go on to other examples; this is just to distinguish names from abbreviations.)

EXERCISE 1.1[C]   Create the first nine rows of the **FIRST** worksheet as described and print out a screen to show the result.

   Creating   For now, the best guides to creating worksheets are the prescriptions, such as the one on page 5. When you are more experienced, you may be able to follow the listings, such as Listing 1.1 (on page 11), or know which parts of a screen image, such as Screen 1.1, show the content that you enter (e.g., text) and which (e.g., numbers) hide formulae that only prescriptions, listings, or schemata reveal. For this chapter and the next, you would do well to follow the prescriptions.

   Printing   Be sure that your name, class, and the date are correctly shown where suggested on the screen. Move the cell pointer to cell **A7** so that the content line shows the formula you have used. Before trying to print, **Save** your worksheet to disk if possible; printing problems frequently cause computer crashes (failures). To print what is on the screen, locate the key that on many keyboards is labeled PrtSc and, while holding down one of the {Shift} keys, press it once. (We shall call it the {PrintScreen} key; on some newer keyboards, it is labeled this way.) When it is pressed, the printer attached to your computer should begin making a printed copy of what you see on the screen. If the printer available to you is provided through a network, you will briefly see the cursor dance over the screen while the print information is scanned; the information will then be queued (put in line) by the network for printing in its turn.

EXERCISE 1.2[P]   Vary the FIRST worksheet a little by using a different formula and by placing the calculation in a different cell on the worksheet. Make a {PrintScreen} hard copy (paper printout) of your work, with the content line showing the formula.

1.2   Putting the Invisible Assistant to Work

The part of the FIRST worksheet we have written so far makes rather weak use of the invisible assistant, who is supposed to do all the repetitive work for you. This is because we have not yet done anything strictly repetitive. To proceed to that, let us make a table showing the values of the sine function every 0.2 radians. Here is the prescription for it:

C9: t
D9: sin t
You have probably done these two already.

C10: 0
Initial t value.

C11: 0.2+C10
This explains to the assistant that you want each successive value of t to be 0.2 radians larger than the previous value.

C11: /Copy C11 *to* **C12**
Here we list, after the colon, not the cell content but a command you should give after locating your cell pointer here (cell C11). There are several ways to give this command, the easiest being to type the keystrokes /c{Enter}{Down}{Enter}. You should now see the value 0.4 shown in cell C12.

C12:
Nothing to enter here, but see from the content line that cell C12 contains essentially the same formula as cell C11, which it was copied from—a formula that adds 0.2 to the value of the cell just above it. This kind of copying can be done wholesale as well as one cell at a time, as we shall soon see.

D10: @sin(c10)
Begin calculating the second column in the table.

D10: /Copy D10 *to* **D11**
Use keystrokes /c{Enter}{Down}{Enter}. Watch the active line to see that with your {Enter} keystrokes you are giving answers to the spreadsheet's "from...?" and "to...?" questions.

D11:

> This cell should now contain the formula `@SIN(C11)` and display the value 0.198669 for sin(0.2).

C11: `/Copy C11..D11` *to* `C12..C60`

> This is large-scale copying. When you type `/c`, the copying process begins. The active line is asking you where to copy from and suggests the current cell, but by a keystroke {Right} you can choose to copy both cells `C11` and `D11` at once. Press {Enter} to declare you have completed pointing out the area you want to copy from. The active line will then ask where you want these cells copied to. Move down one cell, to `C12`, by pressing {Down} once. Then press the period key (`.`), hereafter denoted {Period}, and move down a few rows; {Down 3} will be our shorthand for pressing {Down} three times. You will notice that the highlight has expanded to stretch all the way from cell `C12`, where you "tacked it down" with the {Period}, to the current cell a few rows below. To get fifty steps in our table we need to copy all the way down to row **60**, so move down a bit faster by pressing {PgDn} to move down a "page" or screenful (20 rows) at a time. Stop when the bottom of the highlight rests on cell `C60` and be sure not to include any column other than `C`; then press the {Enter} key to declare that the highlighted area defines the left edge of the area you want to copy to. Your table should now be complete. Figure 1.1 on page 11 illustrates the relation between the two ranges that you highlighted and the resulting table.

The table you have just made is too long to fit on a single screen, but you can use the cursor keys, {PgUp}, and {PgDn} to inspect any part of it.

Listings and Schemata

After worksheet construction has become familiar, you will not need lengthy keystroke-by-keystroke prescriptions for many standard sections of worksheets. The schema and listing found at the caption Schema 1.1 are two alternative, brief ways to specify portions of a worksheet. Each, in a different way, describes the essential logic of the table of sines you have just produced.

Listings are usually located at the ends of chapters and intended not for reading but rather for reference if you are having problems debugging a worksheet. You can compare any suspicious cell formula in your worksheet with the corresponding cell in the listing to see if there are any differences in detail that might be a clue to why your worksheet is not performing satisfactorily. Listings at the end of each chapter are also used to describe the settings that create the graphs accompanying each worksheet. See Section 1.6 at the end of this chapter for examples of such worksheet and graph listings.

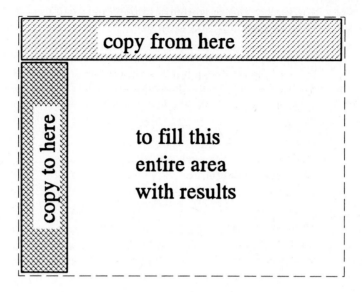

Figure 1.1: Source and target ranges for the /Copy command.

Schemata are representations of the logic in a range of cells, of which the last row is usually copied many times to effect a significant computation in a worksheet. In contrast to a listing, a schema does not use spreadsheet syntax; it names quantities by the sort of mnemonic symbols typically used in mathematics and shows calculations by formulae in standard mathematical notation. It relates to the worksheet by giving a name to each cell and showing each cell in the form *name: content*. It is intended to convey the way you should be thinking about the numbers being calculated. For instance, in the table of sines, we needed a list of angles t_0, t_1, ..., t_n, ..., t_{50} related by the

t	sin t
t_0:	f_0:
0	$\sin t_0$
t_1:	f_1:
$t_0 + 0.2$	$\sin t_1$

```
C9:  't         D9:  'sin t
C10: 0          D10: @SIN(C10)
C11: 0.2+C10    D11: @SIN(C11)
C..:  ...       D..:  ...
```

Schema 1.1: The logic used to tabulate the sine function. Listing 1.1 on the right shows exactly the same information in a standard spreadsheet syntax that the table (schema) on the left shows in a notation more like standard mathematics. The fourth row in each case suggests that the preceding row is to be copied many times.

rule $t_{n+1} = t_n + 0.2$ and a corresponding list of the values of the sine function $f(t) = \sin t$ with $f_n = f(t_n)$. You will learn to think in these terms in designing any worksheet you make or use, and only fleetingly will things like +C11+0.2 and @SIN(C12) run through your mind as your fingers type out the worksheet.

EXERCISE 1.3[C]   Add the tabulation just described to your FIRST worksheet, using the prescriptions in this section. Make a {PrintScreen} copy of the screen as described in Exercise 1.1, showing on the content line a formula from one of the cells in column C. Save the improved worksheet by giving the /File Save command with the keystrokes /fs{Enter}r, which assume that the worksheet already has a filename from having been saved earlier. The r in this command sequence stands for Replace, which is your o.k. to the spreadsheet to overwrite with the updated version your previously saved version of the FIRST worksheet.

EXERCISE 1.4[P]   Modify your FIRST worksheet to tabulate a nonlinear function different from the sine or the cosine. Section B.2 Appendix B lists some of the other functions that are available. Provide a screen dump (use the {PrintScreen} key) with your formula for the new function visible on the content line. You can save this file under the new name FMOD by the command /fsFMOD{Enter} so that it does not overwrite your FIRST worksheet file and will be available for later use.

1.3   Asking the Assistant to Draw a Graph

Tables such as the one you have just made are not very informative because a reader must use a lot of effort to concentrate on a list of numbers and get useful ideas or concepts from them. Graphs communicate more effectively, so we now want to graph the data you have just produced. (If you are returning to work on the FIRST worksheet after doing something else, you will have to retrieve it from your floppy disk, using the /File Retrieve command on page 8.) At the same time we encourage you to get better acquainted with the command-menu structure in your spreadsheet by looking carefully at the active line and the preview line below it as you execute the graphing commands.

The Command Menu

Begin with the spreadsheet in READY mode (which can always be set by pressing {Escape} a few times) and press the slash key / (which in formulae indicates division but here signals the beginning of a command to the invisible assistant). Notice that the active line is now filled with a **menu** of possible commands from which you may choose by typing the first (or key) letter of the one you want. The offerings include File, Copy, and System, which we have met before. You can get an idea of what the result of any choice would be by using the cursor keys to move the highlight in the active line to the command you are considering. The **preview line** (below the active line) will change as this highlight moves and offer information about the highlighted command. If you move the highlight to File, the File menu will be shown on the preview line, including the commands Save and Retrieve, which you have already met, and several others. (In spreadsheets from publishers other than Lotus, the menu names may be slightly different. But consistent with the *de facto* 1-2-3 standard, most are activated by the same key characters that we have used in specifying command keystroke sequences.) Move the highlight to the `Graph` command and note from the preview line that choosing it will lead to another menu.

The Graph Menu

With the main menu (presented in response to the slash key) showing on the active line, press the **g** key to select the `Graph` menu. From this, first select the `Type` menu and its `XY` command to assure that the data associated with the x-axis will have quantitative significance. (Without this specification, most spreadsheets assume a graph called the `Line` type, which expects category labels, such as `Nuts Bolts Screws Nails Clips`, along the x-axis.) Although we have spent time studying the spreadsheet's reactions to our commands, note that to this point we have used only four keystrokes, `/gtx`, to produce the command `/Graph Type XY`.

Pointing to Ranges   After you give the `/Graph Type XY` command you will automatically be returned to the `/Graph` menu, where you must point out what you want to have graphed. The possible data sets (**ranges**) are called `X A B C D E F` on this menu, where `X` will allow you to select the data associated with the x-axis, and the others the data associated with the y-axis. You want to plot your $\sin t$ values as y-axis values in style `A` against the t values along the x-axis. Therefore, select `X` from the menu (by typing the character `x` or `X`); you will be asked to specify the range to use. You may respond either by entering a range specification, here `C10..C60`, or by using the cursor keys to point out the desired range. While the first method may be easier to execute, it is somewhat error prone, so we recommend pointing. To point out this same range, move the cell pointer to the top of the column of t-values, which in Screen 1.1 is cell `C10`. There tack down a corner of the highlight by pressing {Period}. Then, to move rapidly to the bottom of the column, press {End} (on the cursor pad), followed by {Down}; you will jump directly to the bottom of the column of numbers. Seeing that the correct set of data is highlighted, you should then press {Enter} to select it. Use the same procedure, selecting `A`, then pointing out the column of $\sin t$ values, to specify the y-axis data. This completes the basic graph; to see it on your screen, select `View` with a press of the **v** key. If all has gone well, it will look like Figure 1.2 (details are dependent on the graphics resolution of your screen). Press {Enter} to clear the graph and return to the graph menu. The prescription for creating this graph is:

```
C10: /Graph Type XY
 /G: X C10..C60{Enter}
 /G: A D10..D60{Enter}
```

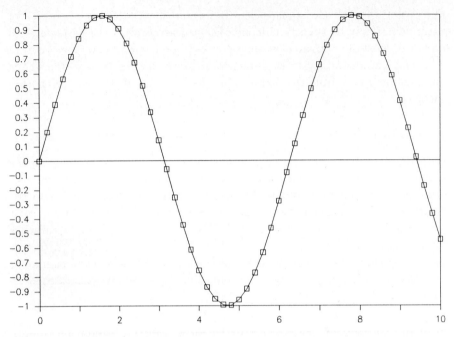

Figure 1.2: Graph produced by your FIRST spreadsheet.

/G: View

    Note that you remain in the graph menu until you explicitly Quit it.

Decorating and Saving Graphs   The other selections on the /Graph menu allow changes in the appearance of the graph; for instance, the keystrokes ocqv, which abbreviate Options Color Quit View, will turn on color (if you have a color monitor) and show you the revised graph. Any unnamed graph specification is lost once the graph settings are changed. To be sure that a satisfactory graph remains available, you should name it before trying more changes. Naming is done with the /Graph Name Create command; then the keystrokes ncFIRST_GRAF{Enter} complete this command. After quitting the /Graph menu and saving your worksheet in satisfactory form, you may wish to return and explore the variations you can design for this graph. {F1}, or {Help}, will be useful to you: whenever this key is pressed, some explanation of the current menu options and their effects will be given. To return from the Help screens to the command you have interrupted, press {Escape}.

EXERCISE 1.5^C Add the graph just described to your FIRST worksheet. After saving your work to disk, make a {PrintScreen} copy of the graphics screen. Because of the potentially higher information content of a bit-mapped graphics screen compared to a text screen, this may take some time. A directly connected dot-matrix printer should begin printing in a few seconds but may take minutes to finish. A laser printer may not show any reaction (apart from perhaps a blinking I/O light) for up to a minute or two, and a network print queue may not allow you to regain control of your computer until after a minute or two. Be patient. But possibly, if this is your first try, your computer may not be properly installed to accomplish this task and may simply have gone dead. It will then need to be restarted, and only the work you have saved to disk will remain. (Section E.1 of Appendix E contains some suggestions for software to make {PrintScreen} work for graphs just as it does for text.)

EXERCISE 1.6[P]   Modify your FIRST worksheet (with its newly defined graph) so that the tabulation begins at $t_0 = 1.0$ and proceeds by the rule $t_{n+1} = 10^{0.02} t_n$. Spreadsheet syntax for x^y is **x^y**; the ^ key is a shifted 6 on most keyboards. Let the function tabulated be $t^{1/3} \sin(6\pi \log_{10} t)$. When you have made the required changes and copied them down their respective columns, press {F10}, or Graph, for an instant display of your new table in graphical form. Does it look like what you expected? When you are satisfied with it, make a {PrintScreen} screen dump to document your result. Then examine and describe the changes in the graph when you change the first value of t (in cell C10) from 1.0 to 10^9 and to 0.001. You might ask a theoretical physicist how these relate to the renormalization group. To save this worksheet so that it does not overwrite either FIRST or FMOD give it a distinct name like FMOD1 when you /File Save it.

EXERCISE 1.7[P]   Retrieve or recreate the worksheet FMOD you produced in Exercise 1.4 and from it create a graph of the function you tabulated. Provide a {PrintScreen} copy of that graph and, to identify the calculation behind the graph, attach a printed copy of the main screen of the worksheet.

1.4   Basic Control of Your Spreadsheet

Although the skills you have acquired while constructing your FIRST worksheet will be sufficient, when practiced and familiar, to let you solve many physics problems on a spreadsheet, you may wish to see them described more systematically and to learn other skills that can make using a spreadsheet easier. The most complete source for such information is the reference manual provided by the publisher of the spreadsheet you use. Here we provide an abbreviated list of spreadsheet facilities most useful for physics applications. You should skim this section now and return to it as a reference when you see a chance to use particular operations it describes. The extensive help facility Lotus 1-2-3 provides (many 1-2-3 clones are less complete in this area) makes our first topic the most important. The later topics serve mainly as suggestions for areas especially useful, with details of their invocation left to the on-line help facility or the spreadsheet's reference manual.

The Help Key   When using 1-2-3 and similar spreadsheets, you can get help just by pressing {F1}, or {Help}. This can be done at almost any time. The spreadsheet will try to give advice relevant to what you are doing at the time. For instance, if you are trying to save a file and the name or directory looks wrong, just press {Help}; the spreadsheet will tell you about its file saving or naming procedures or about how to edit a name it has suggested. The spreadsheet may guess wrong about what you need to know, but it will offer an index of topics (leading often to subtopics) that should let you find something relevant. To select one of these related topics, use the cursor keys to highlight the topic of interest, then hit {Enter}. When you are finished, press {Escape}, and you will return to the task you interrupted.

The Reference Manual   If pressing {Help} does not give you the information you need, refer to the reference section of the manual that accompanies your spreadsheet software. While such manuals are not designed to be read cover to cover, you will probably find that selective reading will not only answer your immediate questions but will also reveal features of your spreadsheet that can make your work considerable easier. To start, you should look for basic skills (cursor control, file handling, /Copy) and study graphing as applicable to XY graphs. After you gain some experience using your spreadsheet you may wish to explore the mathematical functions available, try some data importation, experiment with linear regression, or learn to use matrix operations. You can defer database and macro facilities until you aspire to expertise.

The Escape Key   The {Escape} key, which has no guaranteed position on keyboards but is usually marked Esc, is a convenience and safety feature. It allows you to back up, one step at a time, in any command you have begun to issue and has analogous effects in some other situations. It is always safe to use, except that it will sometimes erase current specifications if you follow it with an {Enter} rather than with several more {Escape}s. By relying on {Escape} to get you out of trouble, you can try unfamiliar commands without studying the reference manuals first. If the menus or the messages on the active line suggest that you are not achieving your aims or if you lack needed information, you can issue several {Escape}s and be returned to READY mode with no harm done. But remember to try {Help} if the situation is merely confusing.

The Control Panel   Lotus calls the top three lines of 1-2-3's screen the **control panel**. You should watch this area of the screen carefully when you issue commands. When you begin to make a cell entry or when you begin to edit the current cell by pressing {F2}, or {Edit}, the second line here, which we have called the active line, will show the tentative new cell entry as you prepare it. The top line, or content line, also shows a mode indicator at its right end. This lets you know what the spreadsheet thinks you are doing; the effects of some keys change accordingly when this indicator changes. For instance, you must be in READY mode for the / key to bring up the main menu. That brings you to MENU mode, where the cursor keys will move a highlight along the menu on the active line. At the same time, brief indications of what the highlighted selection would do (if selected) appear on the preview line below it. A highlighted command may be selected by pressing {Enter}, but we recommend that you normally type the key letter (first or capitalized letter in the command name) to select from a menu. This is faster than moving the highlight for frequently used command sequences, such as **/fs{Enter}r** for the command **/File Save {Enter} Replace** to save a modified worksheet in the same place it was last saved.

Many commands (e.g., **/Copy** or **/File Retrieve**) require additional information to be completed (from ...?, to ...?, which file?). The spreadsheet's invisible assistant will ask for this information in messages appearing on the active line, which usually shows a **default** (best-guess) response. This default can be replaced or edited by using **{Backspace}** and typing in something new, or by selecting from a menu of further suggestions that may appear on the preview line. If you simply type unfamiliar commands from prescriptions without watching their effects step by step in the control panel, you will have a hard time mastering the spreadsheet as a useful tool.

Labels, Numbers, @functions Labels (text for display) always appear on the content line, preceded by a single or double quote or a caret (^). One of these is supplied automatically by the spreadsheet when you enter anything it guesses is text. You may supply this symbol yourself if you wish. The single quote ' causes the label to be left aligned (pushed to the left side) in the cell when extra room is available. The double quote " causes right alignment, and the caret ^ centers the label. Labels too long to fit in a cell are stored (up to 240 characters) and are displayed fully only if they do not overlap information in other cells. Numbers are always displayed right aligned. Large and small numbers—such as $1.234 \times 10^{\pm 56}$— can be entered in the form **1.234E±56**. To force the spreadsheet to interpret a cell entry as a formula, you should begin the formula with a plus sign (or an equal sign in some spreadsheets such as Microsoft Works and Excel) or one of a few other characters that signal mathematics. The **@** sign is one of these. It prefixes the names of built-in functions. You will have use for, at least, **@SIN, @COS, @SQRT, @PI, @LOG,** and probably several others, such as **@SUM, @AVG, @MIN,** etc.

Copying and Pointing The **/Copy** command is central to the use of spreadsheets in physics because short-time changes in physical quantities obey relatively simple laws, and changes over longer periods are just compounded by copying the short step many times. The source and destination of a **/Copy** can be rectangular blocks of cells called ranges, which should not overlap. Normally, several cells in a row are copied to make many similar rows. One copy of the entire source range is made for each cell in the destination range, so in copying part of a row, the destination range is usually part of a column as in figure 1.1 on page 11. Although a range can be specified by typing the locations of two diagonally opposite corner cells in a format such as **C10..E12**, pointing is usually safer if entries already exist in the range because you can see that you are getting the rectangle you want. Pointing becomes available

automatically when the spreadsheet needs you to specify a range. The cursor
keys then move an expandable highlight to cover the rectangle you want. Use
{Period} to tack down one corner of the highlight so that you can control the
opposite corner with the cursor keys. If you press {Period} again, the movable
corner (shown by the small blinking cursor) will jump to successive corners of
the highlight. To free the highlight so that no corner is fixed, press {Escape}
or {Backspace}. Press {Enter} to record the currently highlighted range where
its specification is required.

Spreadsheet Constants   To give correct results, constants in mathematical
formulae must be specially marked during the spreadsheet **/Copy** process. A
formula that simply refers to a neighboring cell will, after copying, become a
formula that refers to the new neighboring cell. This is called relative addressing, which is most often what we want. However, if a referenced cell contains a
constant (mass, time step, velocity of light, etc.), then we want the copied formula always to use the same value from the same cell. This is called absolute
addressing. In 1-2-3 and similar spreadsheets, an absolute reference is marked
by dollar signs. Thus a formula $t_{n+1} = t_n + dt$ to be copied down a column
in a table would be written **C11: +C10+$C$5** if the value of the small dt were
entered as **C5: 0.2**. In schemata, we point out the need for such absolute
references by making constants to be copied boldface, e.g., $t_{n+1} = t_n + \mathbf{dt}$.

Large Cursor Movements   Moving row by row or column by column is
often a slow way to move about the worksheet. When large movements are
desired, there are more efficient ways to get there. Most commonly used is
the two-keystroke command {End}{Down}, which will take you to the end of
a column of formulae, as when you are pointing out a range for graphing. The
{End} prefix works in other directions too and stops at the first change from
filled to unfilled cells or vice versa. {PgUp} and {PgDn} move a screenful up
and down, {Ctrl-Right} and {Ctrl-Left} do the same horizontally, where {Tab}
and {Shift-Tab} also work. {Home} will bring you to the top left corner of
the worksheet. If you remember cell addresses (or the names you have given
them), then the {GoTo} key, {F5}, lets you jump directly to any cell.

The Graph Key   After a graph has been designed, you need not go through
the graph menu (e.g., **/Graph View**) to see it again, even if the data within
its selected ranges have changed. {F10}, also called {Graph}, will recall the
graph directly from **READY** mode. In your **FIRST** worksheet, for instance, you
could change the starting point t_0 to something different from 0 and immediately press {Graph} to see the new values plotted, or you could change the
plotted function in cell **D10**, then copy it down the column (**D10: /Copy**
{Enter} {Period} {End} {Down} {Enter}), press {Graph}, and immediately
see the new function plotted. We will design our worksheets with adjustable

parameters whose effects can be explored by changing the parameters, recalculating if necessary, and pressing {Graph} to view the results.

Recalculating By default (i.e., if you haven't commanded otherwise) the worksheet is recalculated every time you change a cell; changed data take effect immediately. This can be annoying if you are modifying a large worksheet—typical of mechanics problems, for example—where several thousand computations have to be done, requiring a few seconds. The cure for this is to issue the command /**Worksheet Global Recalculation Manual** (keystrokes /**wgrm**) that lets a recalculation be done only on request. Then, if you see the status indicator **CALC** at the bottom of your screen, it is a sign that changes have been made that may invalidate the numbers displayed. In this situation, you should not believe what you see on graphs or elsewhere. To get valid results, you must press {F9}, or {Calc}, to force recalculation, which will remove the **CALC** status warning.

File Handling All your work disappears when you exit the spreadsheet or turn off the computer unless you have saved it to disk using the /**File Save** command, so you should issue this command every ten minutes or so while working, whenever you have added useful improvements to the worksheet. To begin a new worksheet, use /**Worksheet Erase Yes**. To be saved, a newly created worksheet must be given a name. The name cannot be more than eight characters long but may be preceded by a disk drive and/or directory specification, as in **A:PHYSFILE**. The /**File** menu also includes other useful commands. /**File Retrieve** is used to recall a worksheet saved in a previous session or to discard the current worksheet (if you have just made some horrible mess of it) and recall a prior form of it (from the last time you saved it). (When you retrieve a worksheet, the one you currently see on screen is discarded.) The /**File Directory** command allows you to change the disk drive (**A:**, **B:**, **C:**, etc.) and directory (**C:\WBK**, **C:\PHYS101**, etc.) from which /**File Retrieve** offers you a menu of saved worksheets. You may need this command at some point to switch between your own disk and a network drive on which the instructor has made some worksheets available. The /**File List Worksheet** command will show you all the worksheets in the current directory.

If you have laboratory data you want to analyze and compare with theoretical models by using your spreadsheet, those data can be brought into your spreadsheet easily if the measuring equipment is able to write the data in rows and columns in an ASCII file. While such equipment is not assumed for this book, the data transfer is quite simple using the /**File Import Text** and /**Data Parse** commands described in your spreadsheet reference or via {Help}.

Editing Worksheets always get changed. If poor, they get improved. If good, they get adapted to solve similar problems. Fortunately, spreadsheets provide many conveniences to make change possible. The simplest editing is merely to change the content of a cell by entering something different, which then replaces the previous content. This would be inefficient for a small change in a long formula; instead {F2} (called {Edit}) is pressed, and the cell content is displayed on the active line, where {Left}, {Right}, and {Backspace} keys allow character-by-character changes. Press {Enter} to incorporate the changes in the cell or {Escape} to keep the original content.

Several other commands, which you can study when needed from {Help} or a reference manual, allow large-scale changes. They can be used to improve the appearance and readability of a worksheet or to reorganize it when you are adapting it to a different problem. These commands include /Move, /Range Erase, and /Worksheet Insert (or Delete) Row (or Column).

1.5   Your New Spreadsheet Skills in Review

In this chapter you have had a chance briefly to try out the following aspects of using a spreadsheet:

- **Cursor keys** These move you around the worksheet in READY mode and include the four arrow keys, {Up}, {Down}, {Left}, and {Right}, as well as the larger motions {PgUp}, {PgDn}, and {Home} and the prefix {End}.

- **Labels** Text for display is entered into the current cell in READY mode by typing, concluding with {Enter}. It may be preceded by a positioning character, ', ", or ^, one of which is necessary if the text could be mistaken for a formula. The {Backspace} key is available to correct typographical errors.

- **Numbers** Values to be used in computation are entered into the current cell in READY mode by typing, but no spaces or positioning characters are allowed. Finish the entry with {Enter}.

- **Formulae** Enter formulae into the current cell in READY mode as you would numbers, but make the first character a + to avoid confusion with text. Values in other cells are invoked by their cell address (e.g., C12). Standard arithmetic operations to combine them are: +, -, *, /, and ^.

- **Functions** Many built-in functions are available for use in formulae, including exponential, logarithmic, and trigonometric functions. Each has a required syntax and a name that begins with the character @. Pressing {F1}, call {Help}, will lead you to these details.

- **Editing** The previous cell content is discarded whenever a new entry is made. To change a cell without retyping its content completely, use {F2}, called {Edit}, to enter EDIT mode.

- **Main menu** In READY mode, the slash key / brings up the main menu of spreadsheet commands and puts you in MENU mode, where the cursor keys move a highlight among menu items.

- **Copying** The most important command is `/Copy` which allows any row of computations to be repeated many times. (Related topics are Ranges and Pointing.

- **Graphing** The `/Graph` command brings up a menu that allows converting tables of numbers into graphs. The four-character sequence `/gtx` executed from **READY** mode gives the `/Graph Type XY` command to select the style of graph used for plotting functions. Other selections from the graph menu select the data for graphing, and `/Graph Name Create` saves these specifications with the worksheet.

- **Saving to disk** The `/File Save` command writes out the current worksheet from the temporary electronic memory (connected to your video screen) to the semipermanent magnetic memory (imbedded on a floppy—or hard—disk). The `/File Retrieve` command discards the current worksheet in electronic memory and makes current a worksheet previously saved on disk.

- **Ranges** Rectangular blocks of cells are called ranges and are designated by two opposite corners, as in `A1..C7`. They must be specified (either in that notation or by pointing) when you are copying and selecting columns of numbers for graphing.

- **Pointing** When a command or a function needs a range specified, using the cursor keys starts the **POINT** mode, highlighting the proposed range. The period character . (also designated {Period}) is used to tack down a corner of the highlight. Then the cursor keys are used to extend the highlight to the opposite corner. Typing the next character of the command or function accepts the highlighted range.

- **Escaping** {Escape} goes back one step in any operation (such as giving a command) to remove the current choice and look at the prior choices again. In **READY**, mode it does nothing.

- **Help** {F1}, called {Help}, brings up information about your current activity and displays a menu of topics on which you can request further information.

- **Watch the screen** If you enter keystrokes from our prescriptions without carefully watching the effect of each keystroke, you will be making decisions and accepting choices without understanding what you are doing. Never press two keys in quick succession unless the action of the first is so familiar that you know its consequences in your current context.

1.6　Listings

```
A1: 'your name
A2: 'your class & section     D2: 'FIRST, a simple worksheet.
A3: 'today's date             D3: 'Find that you can type, calculate,
                              D4: 'and graph what you want.
A6: 0.5
A7: @PI*A6
            C9:  't           D9:  'sin t
            C10: 0            D10: @SIN(C10)
            C11: 0.2+C10      D11: @SIN(C11)
            C..:  ...         D..:   ...
            C60: 0.2+C59      D60: @SIN(C60)
```

Listing 1.2: Cell entries to make your FIRST worksheet. Note that you do not type in the formulae in cells C12..D12 through C60..D60; use the /Copy command to produce them. When a range of cells is produced by copying, an entry such as C..: ... stands for the cells that are just copies of the line above in the listing and is usually followed by the last line in the copied range.

```
        Name:   FIRST_GRAF   Type:   XY
        Range
            X:   C10..C60
            A:   D10..D60
```

Listing 1.3: Graph settings for the FIRST worksheet.

Chapter 2

Describing Change

In Chapter 1 you learned how to create a worksheet to do repetitive calculations and to make graphs of the results of such calculations. Now you will apply these skills to solving an important class of physics problems involving motion.

2.1 Derivatives Describe and Predict Change

The area of physics that develops a vocabulary for describing motion is called **kinematics**. For the one-dimensional motion that concerns us here, the basic description is a function $x(t)$ that specifies the position of a particle as a function of time. Experiments on this class of motions seek to tabulate this function in simple controlled cases such as a ball being dropped, a mass bouncing as it hangs on the end of a spring, or a glider moving on an (almost frictionless) air track. Theories try to find a clear, communicable, and simple way to predict this $x(t)$ function, i.e., to draw a graph of x versus t that will agree with experimental measurements.

   The great breakthrough on this problem, culminating in the work of Galileo and of Newton, was the discovery that neither position x nor its velocity v had immediate causes, but that acceleration did. This was and is difficult to accept because thinking people have learned to get most of their information through their eyes. The eye is very sensitive to velocity but often ignores acceleration, so we find it hard to accept that the fundamental laws of physics ignore velocity. Our tactile senses (e.g., of touch or equilibrium) do give us the important hint that we need to study acceleration, the second derivative of $x(t)$.

Finite-Difference Methods

On a computer, the numerical treatment of derivatives cannot take the limit $\Delta t \to 0$, a mathematically essential part of the definition of a derivative such as the instantaneous velocity $v = dx/dt = \lim_{\Delta t \to 0} \Delta x/\Delta t$. Also, most computers cannot easily show the Greek delta Δ. Our working rule is to write dt instead of Δt and to let it be a nonzero number small enough to give results (to the accuracy with which we calculate) independent of its exact size. In our computations, then, the velocity will typically be computed as

$$v = \frac{x_{n+1} - x_n}{t_{n+1} - t_n} \qquad (2.1)$$

and we will write

$$dt = t_{n+1} - t_n \qquad (2.2)$$

or

$$v = [x(t + dt) - x(t)]/dt \quad . \qquad (2.3)$$

This is, of course, just what is normally defined as the average velocity over the interval from t_n to t_{n+1}; when we need good identification, we will label this velocity $v_{n+1/2}$. By a similar procedure, we can calculate accelerations numerically from a table of velocities by forming velocity differences and dividing by dt to approximate $a = dv/dt$ in finite-difference computations.

Description versus Prediction

From the experimental viewpoint, one measures $x(t)$ by taking a movie or a strobe-lit photograph or by making a timed spark marker record. Then one uses equation 2.3 to compute the (average) velocity during each interval and the similar equation

$$a = [v(t + dt) - v(t)]/dt \qquad (2.4)$$

to compute the acceleration. (The dt here is the fixed interval between data points, i.e., the time between movie frames, strobe flashes, or sparks.) This procedure gives a description of the motion at the levels of position, velocity, and acceleration. The derivatives $v = dx/dt$ and $a = dv/dt$ are approximated numerically by finite differences. A worksheet that does these calculations is described in Section D.1 in Appendix D. The step from x to v and a is also straightforward if $x(t)$ is proposed as an analytic formula and if you have learned enough calculus to carry out the differentiation of the functions involved.

From the theoretical viewpoint to be developed here, the primary understanding is at the level of accelerations (as produced by forces). It is then necessary to turn the mathematics backward to find the velocity and position by integration. This is a more advanced technique in analytical calculus (especially if the acceleration is given not as a function of time but as a function of position and velocity), but by our numerical methods, it is just as easy as differentiation. To find the velocity from a given acceleration, we have only to rewrite equation 2.4 as

$$v(t + dt) = v(t) + a\,dt \qquad (2.5)$$

so that it gives the new velocity in terms of the acceleration and the previous velocity. We can similarly rewrite equation 2.3 so that it predicts the new position from the velocity and the old position:

$$x(t + dt) = x(t) + v\,dt \qquad (2.6)$$

This approach is developed in a more specific physics example in the next sections.

2.2 Adding Some Physics

How does a ball move if it is thrown straight up? You hardly need physics to answer such a simple question. It first rises to some maximum height, losing speed as it gains altitude, then falls, gaining speed as it drops, until it hits the ground. You do, however, need some physics to go beyond such a qualitative description of the ball's motion. A more detailed answer to the question involves specifying the altitude of the ball above the ground at each instant of time. In the shorthand of mathematics, the position of the ball expressed as a function of time can be denoted by $x(t)$. If $x(t)$ is known, the ball's velocity, $v(t) = dx/dt$, and its acceleration, $a = dv/dt = d^2x/dt^2$, can be calculated. A knowledge of these three kinematic variables then tells you all that you need to know about the ball's motion.

In fact, you have probably already learned that the ball moves, near the earth's surface, with a constant downward acceleration of $9.80\,\text{m/s}^2$, usually denoted by g. While this assumption is valid only when air drag is negligible, most physics texts blithely present many such problems, even though air drag has a very noticeable effect on a real ball's motion. The reason most texts ignore the effects of air drag is, of course, that the presumed equations for the motion of the ball, such as the famous $x = x_0 + v_0t - (1/2)gt^2$, are easily obtained using this simple but inaccurate approximation.

By constructing a worksheet to obtain a numerical solution for the motion of the projectile, including the effects of air drag, we can go beyond the simple case of constant acceleration. To do this, we must first construct a mathematical model for how air drag affects the acceleration of the ball. Then we will work back from this acceleration to find the velocity and position and make graphs of all these kinematic variables.

A Simple Model

The acceleration of a freely-falling object is $9.80\,\text{m/s}^2$ straight down. If we agree to call the upward direction positive, then this acceleration can be represented as $-g$. The effect of air drag is to modify the size (and possibly even the direction) of this downward acceleration, depending on the velocity of the projectile. For example, a ball thrown straight upward slows more quickly than it would if air drag were negligible. In this case, the magnitude of its acceleration is larger than g though the direction is still downward. Conversely, a skydiver in free fall will slow abruptly when the parachute opens because the effects of air drag are larger than those of gravity (which would cause the descent to speed up if air drag were negligible). Since the parachutist's velocity is downward (thus negative) and decreasing, the sudden acceleration is actually upward, in the opposite direction from the acceleration due to gravity alone.

A simple but reasonably accurate model for the effect of air drag on a projectile postulates that the acceleration due to drag alone is proportional to the square of the velocity of the projectile and is in the opposite direction. This model is certainly consistent with what every bicyclist knows: air drag is more significant at high speeds than at low, and it tends to slow you down. Put algebraically, $a_{\text{drag}} = -Cv|v|$, where v is the velocity of the projectile, $|v|$ is its speed, and C is a constant that depends on the size, shape, and mass of the object and on the density of the air. (Since all motion is in one dimension in this chapter, we do not need to treat x, v and a as vectors—signed real numbers suffice.)

Another example offers insight into how to combine the effects of air drag and gravity. Consider what happens when raindrops form at an altitude of 2 km and fall to the ground. Without air drag, they would be moving at some 700 km/h (or about 500 miles per hour) when they near the ground, giving new significance to being caught out in a spring shower. What actually happens, of course, is that air drag increases as the drops fall faster, reducing toward zero their downward acceleration. This limits their velocity to well below that of a truly free-falling object. In fact, as the acceleration of the drops approaches zero, their downward velocity approaches a limiting, constant value, called the terminal velocity, v_{term}.

We now have all the pieces needed to assemble a formula for the acceleration of a projectile moving vertically near the surface of the earth under the combined influence of gravity and air drag. The conditions the acceleration must satisfy are:

- When the velocity is zero, the acceleration must be g downward.

- As the acceleration approaches zero, the velocity must approach v_{term} downward.

- The contribution to the acceleration due to air drag must be proportional to v^2 and in the opposite direction to v.

An expression satisfying all of these conditions is

$$a = -g - g(v/v_{\text{term}})|(v/v_{\text{term}})| \quad . \tag{2.7}$$

It is important to note that you must supply an empirical value for v_{term} in this expression. A more detailed analysis is required to calculate how v_{term} (or the constant C above) depends on the properties of the projectile and the air.

Although we now have a formula for calculating a, we still need to apply equations 2.5 and 2.6 to calculate $x(t)$ and $v(t)$ from $a(t)$ and the initial conditions of our problem. This is the subject of the next section.

$h = h_i + \delta t$

labels:	t	x	v	a
init data:	t_0:	h_0:	v_0:	a_0:
	0	h_{init}	v_{init}	$-\mathbf{g} - \mathbf{g}v_0|v_0|/\mathbf{v}_{\text{term}}^2$
typical	t_1:	h_1:	v_1:	a_1:
line:	$t_0 + \mathbf{dt}$	$h_0 + v_0\,\mathbf{dt}$	$v_0 + a_0\,\mathbf{dt}$	$-\mathbf{g} - \mathbf{g}v_1|v_1|/\mathbf{v}_{\text{term}}^2$

Schema 2.1: The DRAG1 worksheet. From a formula for the acceleration caused by gravity and air drag, this schema calculates the changes in velocity and position implied by this acceleration, solving for the motion $x(t)$ by what is called Euler's method.

2.3 Solving for the Projectile's Motion

It is useful to think of equations 2.5 and 2.6 as "marching orders" telling the projectile what its next velocity and position are to be once it has arrived at its present location.

These marching orders also provide a recipe, termed Euler's method, for calculating an entire set of velocities and positions at a sequence of instants, each separated from its predecessor by the same short time interval dt. The repeated application of these two equations at each successive step will reproduce the motion of the projectile, provided the time interval dt is small enough to approximate the limit appearing in the mathematical definition of the derivative. How small dt must be to produce a reasonable approximation of the exact solution depends on several factors. Among these are the extent to which the acceleration and the velocity change between t and $t + dt$ and, of course, the degree of accuracy desired in the solution. For now, you can take the empirical approach of just trying different values starting with $dt = 0.1\,\text{s}$. Other values of dt will show how sensitive the numerical solution is to this choice. Some exercises at the end of this chapter invite you to explore this issue in greater depth. In Chapter 4 we will return to the question of numerical approximations, with the goal of improving on the simple Euler method we adopt here.

Schema 2.1 summarizes how to construct a worksheet, called DRAG1, using Euler's method to solve for the vertical motion of a projectile. While this schema contains more information than the one you saw in Chapter 1, the rules for reading and interpreting it remain the same. Each row in the schema represents one line of the final worksheet. Within each row, the top line gives the symbolic name of the number to appear in the corresponding cell of the worksheet; the second line gives either the formula contained in the cell (e.g., $h_0 + v_0 dt$) or the name of an initial numerical value to be placed in that cell (e.g., h_{init}). Names printed in bold type, such as $\mathbf{dt}$, refer to absolute (rather than relative) cell addresses, a detail that will be reviewed at the appropriate point in the next section.

2.4   DRAG1: Numerical Solution

You now need to take these physical and mathematical ideas and use them to build a worksheet that models the motion of the projectile. Along the way, you also want to be sure to provide, in the worksheet, information sufficient to identify the problem being solved and who is solving it. Remember, the worksheet not only is a means of modeling a particular physical system; it also is your final report of your work, whether for your instructor (now), for your employer (later), or for yourself (now or later).

With this in mind, load your spreadsheet program and start with a blank worksheet. To begin, enter your name, today's date, your class and section, and a title for your new worksheet:

A1: *your name and today's date*
A2: *your class and section*

D2: DRAG1: air resistance problem
E3: quadratic force law
E4: Euler integration

A5: \-
A5: /Copy A5 *to* B5..H5
   The backslash \ causes the character following it to fill the entire cell. That cell is then copied to create a line separating your title page from the input page that follows it. Recall from Chapter 1, page 17, that the **/Copy** command can be completed in several ways. The best choice here is to accept the source (from...?) cell **A5** by pressing {Enter} and then to highlight the destination (to...?) range by pointing to the remainder of the row visible on your screen.

Next, specify the values of the parameters describing the projectile's initial state and those determining its subsequent motion:

A6: Construction parameters:

B7: "g =
C7: 9.8
D7: m/s^2
   Gravity's contribution to the projectile's acceleration.
   Don't forget your units!

The underscore character _, which is used to suggest subscripts in names below, is a shifted hyphen on most keyboards.

B8: "v_term =
C8: 30
D8: m/s
   A first guess for the terminal velocity.

A9: Initial data:

B10: "h_init =

C10: 0

D10: m

The projectile is launched from ground level.

B11: "v_init =

C11: 60

D11: m/s

Is a man or machine throwing this ball?

A12: Approximation:

B13: "dt =

C13: 0.1

D13: s

You step through the motion at 10 frames per second.

A14: \-

A14: /Copy A14 *to* B14..H14

Separates the input page from the work block that follows it.

Finally you must tell the computer how to calculate the motion of the projectile:

A15: name:

C15: ^time

D15: ^height

E15: ^velocity

F15: ^accel

These four entries serve as labels for the columns of the work area of your worksheet.

A16: units:

C16: ^[s]

D16: ^[m]

E16: ^[m/s]

F16: ^[m/s^2]

And these are the units associated with them.

A17: labels:

C17: ^t

D17: ^h

E17: ^v

F17: ^a

Each of these four labels will be used to name the cell immediately below it.

At this point we will take a small side trip to exploit a useful feature of 1-2-3, namely its ability to associate descriptive names with specific cells in a worksheet. You can, for example, attach the name `G` to cell `C7`, which contains the numerical value of the acceleration due to gravity. Then, when you require this value in a formula, you can type the name `G` instead of the cell address `C7`. Even better, if you do type or point to `C7` while constructing a formula, 1-2-3 replaces `C7` (in the formula) with its name, `G`, as soon as you hit {Enter}. This automatic replacement of addresses with names makes it easier to read and interpret the formulae in your worksheets. Of course, this works only if you take the time to give meaningful names to important cells. You can attach names to some cells in this worksheet now to see how this works in practice.

To attach a name to a cell, the crucial command sequence is `/Range Name Create`. When this command sequence is invoked, 1-2-3 will first ask you for the name you wish to assign. Your may choose a previously defined name (if you wish to redefine it) or you may type in a new name. 1-2-3 will next ask you to specify the cell (or range of cells) to which you wish to apply the name. Position your cursor as indicated and try this:

> `C7: /Range Name Create G{Enter} {Enter}`
> Assigns the name `G` to cell `C7`.

Incidentally, 1-2-3 uses only capital letters for such named ranges, so whether you typed **g** or `G` in the previous line, to 1-2-3 the name is now `G`. For the same reason, you cannot use lower-case letters to distinguish named cells from one another; `v_term`, `V_term`, and `V_TERM` are all the same to 1-2-3. But names can be up to 14 characters long, so you should have no difficulty coming up with unique descriptive names for the important parameters in any problem.

Continue assigning names to important cells in your worksheet as follows:

> `C8:  /Range Name Create v_term{Enter} {Enter}`
> `C10: /Range Name Create h_init{Enter} {Enter}`
> `C11: /Range Name Create v_init{Enter} {Enter}`
> `C13: /Range Name Create dt{Enter} {Enter}`
> `D2:  /Range Name Create title1{Enter} {Enter}`
> `A1:  /Range Name Create title2{Enter} {Enter}`

1-2-3 also provides a way to assign separate names to many parallel cells. For example, take the four variable names in the 17th row of your worksheet and assign each to the cell immediately below it, in the 18th row, as follows:

> `C17: /Range Name Labels Down {Right 3} {Enter}`
> Here {Right 3} means to press the right cursor key three times.

At this point, you may be getting a bit concerned about remembering all the cell names you have just created. Fortunately, 1-2-3 provides a way to create a table right in your worksheet listing all the named cells. To do this, you must first move the cursor to an unused portion of the worksheet and then tell the invisible assistant to generate a complete list of named cells. A prescription for this follows:

> **C17:** {Home} {Tab}
>> This moves the cursor to cell **I1** and positions it at the upper left corner of a blank screen. This area, just to the right of your home block, is a convenient location for notes about the worksheet. The {Tab} key is usually one of the oversized keys on the left side of the keyboard, marked with a pair of heavy arrows.

> **I1:** `Notes page:`
> **I8:** `range names:`
>> These are just two reminder labels.

> **I9:** `/Range Name Table{Enter}`
>> This actually creates the desired table.

Naturally, as you add named cells to your worksheet, you should repeat the last step of this prescription to generate an updated table. Then, whenever you need to remind yourself of the name you chose for a particular variable, the {Home} {Tab} key sequence will carry you right to your notes page.

Now that you have names attached to all the important variables, you can exploit these names in the formulae relating the position, velocity, and acceleration of the projectile. One way of doing so is by typing the appropriate variable names directly into the formulae, as in the next part of this worksheet's prescription. Alternatively, you can point to each cell as needed while constructing your formula. In this case, the cell references will appear in their old form (e.g., **C7**) on the active line near the top of the screen as you build the formula. As soon as you hit {Enter}, however, all references to named cells in your formula will be converted to their assigned names on the contents line, making the formula easier to interpret.

Whichever technique you use in constructing these formulae, you must ensure that references to constants, such as **G** or **DT**, continue to refer to the correct cell when these formulae are copied to fill out the work area of the worksheet. This is done by making such references absolute. (This was briefly discussed in Section 1.4 on page 18.) One way to lock the reference to a particular cell is to type *one* dollar sign before its name (as in **$DT**) or *two* dollar signs in its explicit reference (as in **$C$13**). In references such as **$C$13**, the first dollar sign locks the column reference while the second locks the row reference. In more advanced applications, mixed references are possible in which, say, the row reference is absolute, but the column reference is adjusted when the formula is copied. For now, you will use only full absolute reference such as **$DT** or **$C$13**.

You can also use {F4} to convert a cell reference from relative to absolute (either full or mixed). To do this, you simply hit {F4} after typing the name or reference of the desired cell but before typing the rest of your formula. You can also use {F4} while pointing at a cell whose reference should be absolute. {F4} can even be used to convert cell references from relative to absolute in formulae you are editing (using {Edit} or {F2}). In this case, you merely position the cursor anywhere within or just after the desired cell reference and hit {F4} to convert it.

With these comments out of the way, here are the prescriptions for the formulae needed to solve the motion of the projectile:

A18: `init data:`

C18: `0`
D18: `+H_INIT`
E18: `+V_INIT`
F18: `-$G-$G*(V/$V_TERM)*@ABS(V/$V_TERM)`
> These formulae specify the initial values of the time, height, velocity, and acceleration of the projectile.

A19: `typical row:`

C19: `+T+$DT`
D19: `+H+V*$DT`
E19: `+V+A*$DT`
F19: `-$G-$G*(E19/$V_TERM)*@ABS(E19/$V_TERM)`
> These formulae relate the time, height, velocity, and acceleration now to their values one time step earlier. The formula in cell **F19** can simply be copied from cell **F18**.

A20: `copied rows:`

C19: `/Copy C19..F19` *to* `C20..C118`
> This is where the invisible assistant earns his keep, repeating each of the calculations you specified in cells C19 through F19 ninety-nine times! (For details on how to execute a large scale /Copy command, see the example on page 10.)

If all goes well, you should have a sheet much like the one pictured in Screen 2.1. All that should be missing is the descriptive information in cells E7 through E13, which you can easily add to complete your copy of DRAG1. Schema 2.1 on page 27, along with the listings at the end of this chapter, should help you debug your worksheet in case your numerical results differ from those in Screen 2.1.

Notice that the worksheet you have created is structured for easy interpretation. The top section, or title page, provides basic information about the worksheet. Below it, the input/output (I/O) page pulls together the constuction parameters, initial data, and numerical approximations which define the particular example. This page can also provide space for calculated results as needed. The notes page, off to the right of the part of DRAG1 visible in

```
F18: -$G-$G*(V/$V_TERM)*@ABS(V/$V_TERM)                              EDIT
     -$C$7-$C$7*(E18/$C$8)*@ABS(E18/$C$8)
```

	A	B	C	D	E	F	G	H
1	your name & today's date							
2	your class & section			DRAG1: air resistance problem				
3					quadratic force law				
4					Euler integration				
5	---------	---------	--------	---------	----------	----------	---------	-------
6	Construction parameters:							
7		g =	9.8	m/s^2	Earth surface gravity.				
8		v_term =| 30	m/s	Depends on the object that moves,				
9	Initial data:				and the air it moves through.				
10		h_init =| 0	m	Initial altitude.				
11		v_init =| 60	m/s	Initial velocity.				
12	Approximation:							
13		dt =	0.1	s	Time step in the computation.				
14	---------	---------	--------	---------	----------	----------	---------	-------
15	name:		time	height	velocity	accel		
16	units:		[s]	[m]	[m/s]	[m/s^2]		
17	labels:		t	h	v	a		
18	init data:		0	0	60	-49		
19	typical row:		0.1	6	55.1	-42.8587		
20	copied rows:		0.2	11.51	50.81412	-37.9159		

```
DRAG1.WK1
```

Screen 2.1: The DRAG1 worksheet. Shading (grey wash) is used in this representation of your computer screen to show emphases displayed in other ways on your terminal. The darker grey areas here are often presented in reverse video on computer screens. The lighter grey areas here correspond to cells that are often displayed with high intensity or colored text. This feature does not occur spontaneously; only cells that you designate are displayed in this way. To mark a cell or range of cells for this form of display, use the /Range Unprotect command.

Screen 2.1, allows space to record supporting information such as a table of named ranges. Finally, the work block, immediately below the I/O page, contains the formulae used to solve for the motion of the projectile. While we will refine this block structure in Chapter 3, these basic groupings of information will be used in all our later worksheets.

EXERCISE 2.1[C]   If you have not already done so, return to the start of this section and follow its prescription to construct the worksheet DRAG1. When the results of your worksheet agree with those in Screen 2.1, save your sheet (with the name DRAG1) on your personal floppy disk. Then pick a new value for v_{init} between +50 and −50 m/s and examine the resulting solution for the motion of the projectile. Provide {PrintScreen} copies of the topmost and bottommost screens of your worksheet.

2.5 `DRAG1`: Interpreting Your Results

In one sense, you are finished solving this problem. The worksheet you have created tabulates the position, velocity, and acceleration of the projectile for the first ten seconds of the projectile's flight. If you change any of the input parameters, the invisible assistant immediately recalculates a new solution for you. This is not, however, the end of the road. For you can now use your worksheet to build an intuitive understanding of how such a projectile moves under a wide range of circumstances. You can do this in much the same way you first learned how actual projectiles move, namely by changing the initial conditions and watching the resulting motion. But to see more clearly the motion of your projectile, you need to turn your tabulated results into graphs.

The First Graph

You will begin by plotting the height of the projectile as a function of elapsed time. The procedure is similar to the one you used in Chapter 1 to make a graph of $\sin(t)$ versus t. In brief, you must go to the `Graph` menu, select the `XY` graph type, highlight the ranges you wish plotted, add labels to be sure the meaning of your graph is clear, view the graph as you make changes to be sure they are what you want, and finally give the graph a name for future use.

Reload `DRAG1` and follow this prescription to add a graph to your worksheet:

> C18: /Graph Type XY X {Period}{End}{Down}{Enter}

This key sequence selects the XY type of graph and tells 1-2-3 that you want to plot time along the horizontal axis. {Period} tacks down the highlight to cell `C18`; then the {End} {Down} {Enter} sequence extends the highlight to select (as the horizontal axis or `X` range) the time values `C18..C118`.

The preceding command left the cursor on cell `C18` with the `Graph` menu still active. This is indicated by the way we prescribe the next step:

> /G C18: A {Right} {Period}{End}{Down}{Enter}

This tells the computer which range to plot along the vertical axis. If you now select `View` (from the still-active `Graph` menu) by typing `v`, you should see an unlabeled plot of h versus t. When you've seen enough of your graph, a tap on any key will bring you back to the display of your worksheet.

You now need to add labels to your graph's axes as well as titles indicating the name of the problem being solved and the identity of its authors. The tools to do this are tucked away in the `/Graph Options Titles` menu. Using these tools, you can attach the label `time [s] -->` to the x-axis by starting from the `Graph` menu (which is still active) and following this prescription:

> /G: Options Titles X-Axis time [s] -->{Enter}

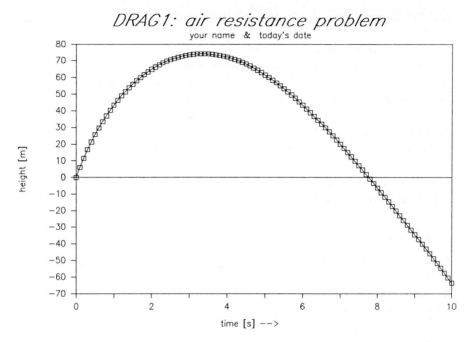

Figure 2.1: A graph of projectile height versus time from DRAG1.

This will label the *x*-axis and leave you at the **Graph Options** menu. (The **/G** in place of the usual cell reference is a reminder that the **Graph** menu is active; the position of the cell pointer is irrelevant and is therefore omitted.) From this point, you can use the **Titles** menu selection three more times to finish labeling and titling your graph.

```
/GO: Titles Y-Axis height [m]{Enter}
/GO: Titles First \title1{Enter}
/GO: Titles Second \title2{Enter}
```

Here, the **/GO** indicates that the **Graph Option** menu is active for the commands that follow. If you now select **Quit** from the **Graph Options** menu, you will arrive back at the **Graph** menu where the **View** option should present you with a fully labeled and titled graph similar to that shown in Figure 2.1.

When you examine this graph, you may be surprised to see that the two lines of the title at the top of your graph are not what you typed. Rather, they are the contents of the cells named **title1** and **title2**. Such indirect labeling has the advantage that any change in these referenced cells will automatically be reflected in the title of this and similar graphs. As you can see from the prescription above, this handy trick was accomplished by entering **\title1** and **\title2** (i.e., each cell name preceded by a backslash) in place of explicit text. On the other hand, you entered explicit text for the horizontal and vertical axes, since these labels are likely to be different on each graph you define.

Before changing this graph, take a minute now to preserve it in its present form for use in the future. Surprisingly, to do this, the **Save** command in the **Graph** menu is *not* the option you want. (The **Save** command copies the current graph into a separate file from which hard copy can later be produced on plotters or printers.) What you want to keep are the specifications for several graphs actively associated with your worksheet. To do this, all that is required is to give each graph a unique name before work is begun on another. The following prescription will give the name **H_T** to your graph:

```
/G: Name Create h_t{Enter}
```

A More Useful Graph

To visualize its motion better, you can plot the projectile's velocity and acceleration on the same graph as you plotted its height. This is done simply by telling your invisible assistant which columns of numbers are to become the graph ranges **B** and **C**. Since the **Graph** menu is still active with the cell pointer still on cell **C18**, the prescription for this is:

```
/G C18: B {Right 2} {Period} {End}{Down} {Enter}
```

which defines the B-range as the projectile's velocity while

```
/G C18: C {Right 3} {Period} {End}{Down} {Enter}
```

defines the C-range as its acceleration. Selecting **View** will now display the new graph, with all three sets of dependent variables plotted as a function of time.

Before you do anything else, you should relabel the graph to reflect the new information it presents. Two changes are needed: you must remove the label from the vertical axis and add a legend to identify each curve with its appropriate plotting symbol. Here is a prescription to do this:

```
 /G: Option Titles Y-Axis {Escape}
          h, v, a [SI units]{Enter}
/GO: Legend A height{Enter}
/GO: Legend B velocity{Enter}
/GO: Legend C acceleration{Enter}
/GO: Quit
 /G: View
```

If all goes well with this prescription, your new graph is now displayed on your computer's screen. Again, hitting any key will bring you back to the **Graph** menu in your worksheet.

While you are at it, you can improve the appearance of your graph (and learn another feature of 1-2-3) by removing the plotting symbols from one of the three curves. This will leave behind a smooth curve that can still be readily distinguished from the other two. The prescription to do this and display the resulting graph is:

/G: `Option Format B Lines Quit Quit View`

Finally, to give this new graph a name, `HVA_T`, return to the `Graph` menu and execute:

/G: `Name Create hva_t{Enter}`

With this, you have finished constructing `DRAG1`. You may inspect either graph with the command `/Graph Name Use` which offers a menu of all the graphs you have previously named.

It is important to understand the relation between these two named graphs and the third, current graph associated with this worksheet. Since you have just given the name `HVA_T` to the current graph, it is identical to the current graph—for now. If you then make some additional changes to the current graph, these changes will only affect the current graph and not either of the previously saved graphs. While this may seem obvious, it has a very important implication. If you now view either `HVA_T` or `H_T` by executing a `/Graph Name Use` command before first using `/Graph Name Create` to preserve your most recent changes to the current graph, you would lose all the improvements you had made to the current graph. The reason is that `/Graph Name Use` wipes out the current graph's definition and replaces it with that of the named graph specified. So, if you like your current graph, be sure to preserve its definition with a `/Graph Name Create` before going on to do something else.

Before using this worksheet to explore the physics of the projectile's motion, you should save your work on your personal disk. To do this, `Quit` the `Graph` menu and use the `/File Save` command to replace your old version of `DRAG1` with the current version. Remember that the current version has definitions of two named graphs built into it. Since graph definitions are saved along with the rest of the worksheet, there is no need to save the graphs separately. These graphs will simply be there, ready to use, the next time you `Retrieve` this worksheet from your personal disk. (In fact, three graph definitions are saved with this worksheet since the definition of the current graph is saved along with the definitions of any named graphs.)

2.6   Testing the Limits

Once a mathematical model for a physical system is created, we must verify that the model is an appropriate description of the system. An important technique for doing this involves looking at the results the model gives in cases where the answers are already reasonably well known. For example, you probably already know a bit about vertical motion near the earth's surface when air drag is negligible. In this case, the acceleration should be a constant $9.8 \, \text{m/s}^2$ downward while the velocity should be a linear function of time with a constant negative slope, and the height plotted versus time should be an inverted parabola. Does `DRAG1` produce these results if you make the air drag

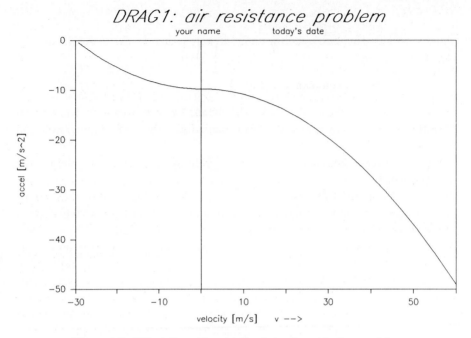

Figure 2.2: This **A_V** graph uses the data shown in Screen 2.1.

small? To answer this question, you must change the numerical value of the construction parameter governing the air resistance acting on the projectile. This parameter is v_{term}, the projectile's terminal (or equilibrium) velocity. In thinner air, the drag on the projectile is less, so its terminal velocity is larger. To reduce the importance of air drag, you must increase the value of v_{term}. The only question is how large a value of v_{term} is large enough to make the air resistance negligible. You will take an experimental approach to answering this and other questions in the course of the following exercises.

EXERCISE 2.2[C]   If you have not already done so, follow the prescription given in this section to create the two graphs named **H_T** and **HVA_T**. When you are finished, save this improved version of **DRAG1** by **Quitting** the **Graph** menu and using the **/File Save** command to update **DRAG1** on your personal disk. Then use the **/Graph Name Use** command to display each of the two graphs in turn and make a {PrintScreen} printout of each graph.

EXERCISE 2.3[P]   Add a third graph to **DRAG1**, plotting the projectile's acceleration as a function of its velocity. Be sure to label each axis of your graph with the variable plotted and its units. Your finished graph should resemble Figure 2.2. Following our conventions for naming graphs, give this graph the name **A_V**, then save this latest version of **DRAG1** on your personal disk. Make a {PrintScreen} printout of the **A_V** graph and then write a paragraph discussing its shape. In particular, state which end of the curve corresponds to the early phase and which to the late phase of the projectile's motion. Comment on why the curve is flat near $v = 0 \, \text{m/s}$, then approaches $a = 0 \, \text{m/s}^2$ as v becomes more negative.

EXERCISE 2.4^C Change the value of v_{term} to 10,000 m/s in DRAG1 and then examine the HVA_T graph to see if this value of v_{term} is sufficiently large to produce the motion expected when drag is negligible. Try other values of v_{term} to find the smallest value that, in the HVA_T graph, yields a difference just distinguishable from the extreme case of negligible drag.

EXERCISE 2.5^E Set the value of v_{term} to 10,000 m/s in DRAG1 as you did in Exercise 2.4. Now examine the A_V graph. Explain why it looks so similar to Figure 2.2 even though the terminal speed is some three hundred times larger now. Does this graph support or contradict your conclusion for Exercise 2.4?

EXERCISE 2.6^E At the other extreme from the case of negligible drag is the case of high drag, in which the terminal speed is, say, 10 m/s. An example might be the vertical motion of a rock dropped into a lake. To see how such projectile motion differs from what you have already examined, set $v_{term} = 10$ m/s in DRAG1 and compare each of your graphs to those for which $v_{term} = 30$ m/s.

EXERCISE 2.7^E It is often useful to characterize a process evolving toward equilibrium by calculating the time it takes to cover some definite fraction of the distance to equilibrium. For example, you can use DRAG1 to determine when a projectile with an initial velocity of zero will reach two-thirds of its terminal velocity. Try this for $v_{term} = 15$, 30, and 60 m/s, noting the two-thirds time in each case. Use your results to comment on how this characteristic time depends on v_{term}. While you are at it, determine the two-thirds time for $v_{term} = 60$ m/s on a planet where the acceleration due to gravity is twice as large as that on earth. That the ratio v_{term}/g has units of time may help you justify your conclusion.

EXERCISE 2.8^E Convince yourself that the time step $dt = 0.1$ s is small enough (see Section 2.3) by verifying that solutions calculated with smaller step sizes yield essentially identical results. Next, determine how large the time step can be before the solution is not equivalent to that obtained with a 0.1 s step. For all of these trials, keep $v_{term} = 30$ m/s. Based on your results here and in Exercise 2.7, formulate a rule of thumb for choosing a safe value of the time step for a given terminal velocity and gravitational acceleration.

2.7   Your New Spreadsheet Skills in Review

- Naming cells with /Range Name Create ....

- Naming cells using adjacent labels with /Range Name Labels Down ....

- Distinguishing between relative and absolute addressing in formulae; using {F4} to change between them.

- Grouping like items together for clarity with:

    - basic information on your title page;

    - important problem parameters on your input/output (I/O) page;

    - additional information, such as a table of range names, on your notes page;

    − the solution of the problem in your work block.

- Using **/Graph Options** ... to control the appearance of your graphs.

- Using labels in your worksheet as titles on your graphs by entering a backslash (\) followed by the name of the cell containing the label.

- Using **/Graph Name Create** and **/Graph Name Use** to store and access several different graph specifications in a worksheet.

2.8   Listings

```
A5: \-  ...  H5: \-
A6: U 'Construction parameters:            I1: 'Notes page:
   B7:  "g =         C7:  U 9.8    D7:  'm/s^2 I8: U 'range names:
   B8:  "v_term =    C8:  U 30     D8:  'm/s
A8: U 'Initial data:                       G      C7      V_TERM C8
   B10: "h_init =    C10: U 0      D10: 'm  H_INIT C10     V_INIT C11
   B11: "v_init =    C11: U 60     D11: 'm/s DT    C13
A12: U 'Approximation:                      T     C18      H      D18
   B13: "dt =        C13: U 0.1    D13: 's  V      E18      A      F18
A14: \-  ...  H14: \-                      TITLE1 D2      TITLE2 A1

A15: 'name:       C15: ^time      D15: ^height
A16: 'units:      C16: ^[s]       D16: ^[m]
A17: 'labels:     C17: ^t         D17: ^h
A18: 'init data:  C18: 0          D18: +H_INIT
A19: 'typical row:C19: +T+$DT     D19: +H+V*$DT
A20: 'copied rows:C20: +C19+$DT   D20: +D19+E19*$DT
                  C..:   ...      D..:   ...
                  C118: +C117+$DT D118: +D117+E117*$DT

      E15: ^velocity      F15: ^accel
      E16: ^[m/s]         F16: ^[m/s^2]
      E17: ^v             F17: ^a
      E18: +V_INIT        F18: -$G-$G*(V/$V_TERM)*@ABS(V/$V_TERM)
      E19: +V+A*$DT       F19: -$G-$G*(E19/$V_TERM)*@ABS(E19/$V_TERM)
      E20: +E19+F19*$DT   F20: -$G-$G*(E20/$V_TERM)*@ABS(E20/$V_TERM)
      E..:   ...          F..:   ...
      E118: +E117+F117*$DT F118: -$G-$G*(E118/$V_TERM)*@ABS(E118/$V_TERM)
```

Listing 2.1: The DRAG1 worksheet. Some cells here show a U prefix in addition to the content that you type. It indicates emphasis (color or intensity) given to that cell by (optional) use of the **Range Unprotect** command in order to improve the appearance of the screen. Cells marked in this way are indicated by the lighter grey wash in screens such as Screen 2.1.

Name: `H_T`   *Type:* **XY**
Titles:
  1st: `\title1`   *2nd:* `\title2`
X-axis: `time [s]   t -->`
Y-axis: `height [m]`

Range   *Format*   *Legend*
  X: `C18..C118`
  A: `D18..D118  Both`

Name: `HVA_T`   *Type:* **XY**
Titles:
  1st: `\title1`   *2nd:* `\title2`
X-axis: `time [s]   t -->`
Y-axis: `h, v, a [SI units]`

Range   *Format*   *Legend*
  X: `C18..C118`
  A: `D18..D118  Both   height`
  B: `E18..E118  Lines  velocity`
  C: `F18..F118  Both   acceleration`

Name: `A_V`   *Type:* **XY**
Titles:
  1st: `\title1`   *2nd:* `\title2`
X-axis: `velocity [m/s]   v -->`
Y-axis: `accel [m/s^2]`

Range   *Format*   *Legend*
  X: `E18..E118`
  C: `F18..F118  Lines`

Listing 2.2: Graph settings for the `DRAG1` worksheet.

Chapter 3

Two-Dimensional Motion

The new physics for two or three dimensions lets each component of acceleration, a_x, a_y, and a_z control the changes in the corresponding velocity, while each velocity v_x, v_y, v_z controls changes in position x, y, z. Thus equations 2.5 and 2.6 are rewritten in parallel for the x and y components:

$$
\begin{aligned}
v_x(t+dt) &= v_x(t) + a_x\,dt & , & \quad v_y(t+dt) &= v_y(t) + a_y\,dt\ , \\
x(t+dt) &= x(t) + v_x\,dt & , & \quad y(t+dt) &= y(t) + v_y\,dt\ .
\end{aligned}
\tag{3.1}
$$

A z component would be treated similarly. Although it might appear that these parallel sets of equations could be treated, one set at a time, in separate worksheets, they are tightly coupled in most applications by the rules for computing the accelerations, each of which may depend on all components x, y, z, v_x, v_y, v_z.

The example developed in this chapter is again a projectile moving under the influence of gravity and air resistance, with air resistance effects proportional to the velocity squared. Our idealization is summarized in a (vector) equation for the acceleration:

$$
\mathbf{a} = \mathbf{g} - \mathbf{v}|\mathbf{v}|g/v_{\text{term}}^2
\tag{3.2}
$$

which has a zero acceleration solution $\mathbf{v} = \mathbf{g}v_{\text{term}}/g$ where $\mathbf{g}$ is a vector pointing downward and $g \equiv |\mathbf{g}|$ is its magnitude. Similarly, $|\mathbf{v}| = v$ is the projectile's speed, the magnitude of its velocity. For use in equations 3.1 the component form of this equation is

$$
\begin{aligned}
a_x &= -v_x\sqrt{v_x^2 + v_y^2}\,g/v_{\text{term}}^2 \\
a_y &= -g - v_y\sqrt{v_x^2 + v_y^2}\,g/v_{\text{term}}^2
\end{aligned}
\tag{3.3}
$$

where the y-axis points upward and the x-axis is horizontal.

Since the kinematics of two-dimensional motion is so similar to that of one-dimensional motion, you can save some effort by starting with DRAG1 and modifying it to include equations for both the vertical and the horizontal components of the projectile's motion. In broad outline, you will need to load `DRAG1`, clear some space to display the additional inputs and results, edit the existing labels to match the new kinematics, make an additional copy of the original columns (tabulating position, velocity, and acceleration), rename the important variables, and then edit the formulae to match the new physics.

3.1   Worksheet Organization–Block Structure

Before focusing on the physics of these improvements to our worksheet, we present some spreadsheet programming skills that can make any worksheet you write easier to adapt to new problems.

A major advantage of using a computer (for writing as well as for calculating) is that work can be reused; it can be copied and then modified to apply to a somewhat different situation. This happens so often that you will fail to take good advantage of computers if you do not plan your work this way. That means organizing work so it can be carried into new areas easily and reliably. An example is our insistence that common parts of graph titles be cells from the title page of the worksheet, rather than being entered directly as the corresponding title for each graph. This constrains the variety of titles a bit in the original worksheet, but it prevents leaving a lot of incorrect titles that you neglect to change when the worksheet is modified. The art of software engineering studies similar guidelines for large-scale projects.

The problem we now want to avoid is that of disrupting one well designed area of a worksheet while improving another. The solution is to give your worksheets a block structure as in Schema 3.1. Then insertions or deletions of either rows or columns in one block do not affect the other block. More complicated worksheets can be made up of several blocks instead of just two. To do your income taxes, let each block contain the lines you use from just one IRS form. Next year add or delete the lines that change and enter the new numbers. Use formulae to carry amounts from one form to another, e.g., from Schedule A to Form 1040, or from 1040 to your state tax forms, so no

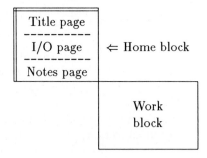

Schema 3.1: Worksheet structured as noninterfering blocks.

number need be entered twice. Add extra columns to the standard forms to do background calculations, as in adding items from several sources.

Let us now put the **DRAG1** worksheet into this block structured form. The top screen of the sheet is shown in Screen 2.1 on page 33. We need to move the work block of the sheet to a region where there is no information either above or to the left of it so that adding rows or columns to the work block will not disturb other areas. Begin by loading your version of **DRAG1** from your personal disk. Restructure **DRAG1** according to the following prescription:

A15: /Move {Right 2} {End}{Right} {End}{Down} {Enter}
{Home} {Tab} {PgDn 2} {Enter}

These are the pointing keystrokes that do a **/Move A15..F118** *to* **I41** if your worksheet is identical to that of Listing 2.1. They would also work if yours is longer than ours (or they would at least show, as you attempted them, that the correct block was not being moved). They should relocate the work block of **DRAG1** one screen over and two screens down from the home screen.

I41: /Range Name Create #{Enter} {Enter}

This gives the name **#** to the upper left corner of the work block for reasons that will be explained.

−−: {Home} {Tab}
I1: /Move {PgDn} {Tab} {Enter} a15{Enter} {Home}
A16: /Worksheet Delete Row A16..A20{Enter}
A18: /Range Name Table {Enter}

We move the notes page that was to the right of the home screen out of harm's way, delete some extraneous blank rows, and rewrite the table of cell names to reflect the changes. The symbol **−−** means that it does not matter which cell is highlighted when you execute the indicated commands.

If you have followed these directions carefully, your sheet should now have two noninterfering blocks, the home block, beginning at the home cell, **A1**, and the work block, containing all our computations, beginning at cell **I36**. The home block is divided (by lines drawn with repeated hyphens) into a title page consisting of a few lines of identification labels, an I/O or input/output page which here contains only input data, and the notes page which you have just moved to this new position.

Now that our work block is safely out of the way, we have a new problem. With the work block no longer visible from the home position, how can we quickly move to the work block to view or modify it. We solved this small problem with another spreadsheet trick by giving the name **#** to the upper left corner of the work block. Why such an odd name? It is easier to show you than to tell you:

−−: {F5} {F3} {Enter}

```
D2: 'DRAG2: 2-dim air resistance problem                              NAMES
Enter address to go to:
#              AX              AY              DT          G
         A      B       C       D       E       F       G       H
1   your name & today's date
2   your class & section       DRAG2: 2-dim air resistance problem
3                                      quadratic force law
4                                      Euler integration
5   --------------------------------------------------------------------
6   Construction parameters:
7              g       9.8 m/s^2
8           v_term      30 m/s
9   Initial data:
10        x_init        0 m          y_init        0 m
11       vx_init       20 m/s       vy_init       45 m/s
12  Approximation:
13            dt      0.1 s
```

```
         J       K       L       M       N       O       P       Q
36    time    x-pos'n  x-vel   x-accel  height  y-vel   y-accel  v mag
37    [s]     [m]      [m/s]   [m/s^2]  [m]     [m/s]   [m/s^2]  [m/s]
38    t       x        vx      ax       y       vy      ay       |v|
39    0       0        20     -10.7243  0       45     -33.9297 49.24428
40    0.1     2        18.92756 -9.42082 4.5    41.60702 -30.5090 45.70993
41    0.2     3.892756 17.98548 -8.33203 8.660702 38.55612 -27.6616 42.54470
```

Screen 3.1: The DRAG2 worksheet.

From anywhere in your sheet this quick key sequence brings you to the head of the work block. The result will make more sense when you remember that {F5} means "go to" and is sometimes written {GoTo}, {F3} can be written {Name} and presents a menu of named cells as possible destinations for the {GoTo} command. This is shown on the preview line above the column letters in Screen 3.1. Because the symbol # comes before all other letters and numbers in the computer's alphabet, it can be selected by pressing {Enter}. If you need reminders of the key functions, remember that help is always as close as your {F1} or {Help} key. In this case, the place to look is under Function Keys in the Help Index.

EXERCISE 3.1[C]   Take a few minutes now to be sure that your sheet is in this new format and that you understand the commands we used to reshape it.

This blocked form of DRAG1 will be a starting point for other worksheets in later chapters, so you should save it now on your personal disk. The sequence /File Save {Enter} Replace will replace the old version of DRAG1 with its improved version.

3.2   DRAG2: Modifying a Worksheet

This section emphasizes spreadsheet editing. You will modify DRAG1 so it can handle two-dimensional motion. If you took a break after putting it into block form, start by reloading DRAG1 into your spreadsheet program. Use the following prescription and Screen 3.1 to guide you through the necessary changes:

D2: {Edit}

Use {Edit} ({F2}) to change the contents of cell D2 to match those shown in Screen 3.1.

E7: /Range Erase E7..E13{Enter}

B10: /Copy B10..D11 *to* E10

--: {GoTo} {Name} #

The last of these steps, {F5} {F3} {Enter}, moves you to the first screen of the work block. The erasure cleared some descriptive information you will have copied from Screen 2.1.

I39: initial:

I40: typical:

I41: copied:

J36: /Worksheet Delete Column J36..J36{Enter}

K36: /Copy K36..M139 *to* N36

By shortening the labels in column I and deleting the empty column J you can view more of the work block on one screen.

With these changes, your work block and your home block are both starting to look more like the corresponding sections of Screen 3.1.

Next, you should position the cursor at each cell containing a label that needs to be revised. Use Screen 3.1 to determine which cells need to be changed. The use of {Edit} and editing commands can save you a lot of retyping here. Remember to change the initial values of the velocity components in cells C11 and F11. Note that all the final equal signs in the labels on the input/output page have been removed to simplify assigning names to the values of these parameters later. You will also need to add a column to your work block to tabulate the speed of the projectile, $v = |\mathbf{v}| = \sqrt{v_x^2 + v_y^2}$. The prescription for this is:

Q36: ^v mag

Q37: ^[m/s]

Q38: ^|v|

The absolute value or vertical bar symbol is found on the same key that produces the backslash character \. On the keycap it may show as a divided vertical bar.

Do not worry that your numerical results do not yet agree with those of Screen 3.1; you still need to correct the formulae relating the initial data to the horizontal and vertical acceleration.

J36: /Range Name Reset

This command discards all the previous cell names and gives you a completely fresh start at giving appropriate names to the cells in DRAG2.

J36: `/Range Name Create #{Enter} {Enter}`
This reestablishes the use of {F5} {F3} {Enter} to move to your work block quickly.

J38: `/Range Name Labels Down J38..Q38{Enter}`
This names cells in the work block. (Before executing this command, be sure row **38** has been revised to reflect the current two-dimensional problem.)

B7: `/Range Name Labels Right B7..B13{Enter}`
E10: `/Range Name Labels Right E10..E11{Enter}`
The reason you removed the equal signs from the labels in cells `B7..B13` and `E10..E11` was to allow this convenient assignment of names to many cells at once.

D2: `/Range Name Create title1{Enter} {Enter}`
A1: `/Range Name Create title2{Enter} {Enter}`
Naming these cells in this way will allow automatic labeling of all of your graphs, as in DRAG1.

A18: `/Range Erase {Right} {End}{Down} {Enter}`
A18: `/Range Name Table {Enter}`
This creates a new reference table of range names.

Q39: `+(vx*vx+vy*vy)^0.5`
Q39: `/Copy Q39` *to* `Q40..Q139`
M39: `-$g*vx*|v|/$v_term^2`
M39: `/Copy M39` *to* `M40..M139`
P39: `-$g-$g*vy*|v|/$v_term^2`
P39: `/Copy P39` *to* `P40..P139`

With the installation of these formulae, your numerical results should agree with those of Screen 3.1. If they do not, check your entries against Listing 3.1 at the end of this chapter and make the necessary corrections. When all is well, save your worksheet (with the name DRAG2) to your personal disk.

3.3 Printing Worksheet Formulae

We have been printing the results of calculations, either as tables or graphs, by using {PrintScreen}. This is convenient and records what we see on the screen, but will not, for instance, print a long table or a few sample rows. The command `/Print` on the main menu offers more control over what gets printed. The following prescription shows how you could, for instance, print just the range-name table from the worksheet.

> **A17:** `/Print Printer`
> `/PP` **A17:** `Range {Period} {End}{Down} {Right} {Enter}`
> `/PP` **A17:** `Go Quit`
>
> When executing these commands, you should see the desired range highlighted before you confirm it with `{Enter}`. You can also add blank lines on the printed page with the **Line** command before you actually print the specified range with **Go**. The **/PP** prefix is a reminder that the **Print Printer** menu is active.

In subsequent exercises you will often be asked to print out a row or two of formulae to show that you have put the proper physics into a computation. If you simply specify the appropriate range and print it with the basic **/Print** menu commands, the result will be the same numbers you see on the screen. To produce a listing of the underlying formulae you must specify appropriate **Options** from the **/Print** menu before you issue **Go**.

> **J38:** `/Print Printer`
> `/PP` **J38:** `Range {Backspace} {Period} {End}{Right} {Down 3} {Enter}`
> `/PP` **J38:** `Options Other Cell-formulas Quit`
> `/PP` **J38:** `Go Quit`
>
> Because you had previously chosen a print range, that range is suggested (by the invisible assistant) in case you want to use it again. The backspace key is used to reject that suggestion and return to your current location.

The **Cell-formulas** option that you choose here remains in effect until you change it with **/PP Options Other As-displayed**.

EXERCISE 3.2[C]   As prescribed, use the **/Print Printer** menu to make a printed range-name table for the DRAG2 worksheet in the **As-displayed** format and a printed listing of the beginnings of the work block (the labels row plus the first three computation rows) in **Cell-formulas** format.

EXERCISE 3.3[E]   Listings in more legible format can be produced with some effort. If you were to produce an original worksheet for a term project, you might want to make a more readable listing than the one provided by the simple **Cell-formulas** format. Try the following procedure: Use **/Print File** instead of **/Print Printer** and supply a filename such as DRAGLIST. The spreadsheet will interpret this as DRAGLIST.prn and send the information there instead of to the printer. You can then reorganize this file with a word processor or an editor to improve its appearance. But you can also use the spreadsheet to do the editing. Use `{End}{Home}` to find the end of the work block, then move right and down to start a new block, name it % or SCRATCH. Then with **/File Import Text** read in your listing. Next put it into row-column format by using the **/Range Transpose** command to convert each vertical list of formulae for a single row into a horizontal (row) list and to delete the vertical duplicates. Then set the column widths with **/Worksheet Column Set-width** `{Right}` until the formulae are fully displayed. When everything looks right on the screen, print this reworked listing in **As-displayed** format to the real printer. To avoid a lot of waste paper, you may wish to limit the page length from the **/PP**

Options menu to 11 lines (instead of 66) and set the top and bottom margins to zero so that the columns that don't fit horizontally are printed below the already printed columns. Or you could /Move blocks of the listing under each other to fit within the width of your printer paper (usually 80 columns) before printing.

3.4 ♠ DRAG2: Graphing Two-Dimensional Motion

In Section 2.5 of the previous chapter you learned how to make graphs that helped you interpret the numerical results contained in a worksheet. With both horizontal and vertical motion to watch, graphs are even more important to conveying an intuitive feeling for this motion.

With all the changes that you made while converting DRAG1 into DRAG2, the old named graphs are now all but useless. For this reason, we will begin by clearing away all the previous graph definitions before defining a new set of graphs. To do this simply follow this prescription:

> ```
> ——: /Graph Name Reset
> /G: Reset Graph Quit
> ```
> The first of these command sequences clears out all the old named graphs. The second clears the definition of the current graph and returns you to the **Ready** mode.

There are two broad classes of graphs that you can use to portray the motion of a projectile in two dimensions. The first of these displays one or more kinematical variables as a function of time. An example would be a plot of speed of the projectile, $|\mathbf{v}|$, versus the elapsed time, t. The second class encompasses parametric plots, which display how one quantity correlates with another rather than how each depends on time. A plot of the projectile's trajectory (height, y, versus horizontal position, x) is a familiar example of such a parametric plot. The following exercises lead you through making and interpreting examples of each of these types of graphs.

EXERCISE 3.4[C]   The most basic graph you can make to visualize the projectile's motion is a plot of its height above the ground versus the horizontal distance it has traveled. The following prescription will create, label, and give the name Y_X to a graph of the projectile's trajectory:

```
K39: /Graph Type XY X K39..K139 {Enter}
 /G: A N39..N139
 /G: Options Titles X-axis x-position [m]{Enter}
/GO: Titles Y-Axis height [m]{Enter}
/GO: Titles First \title1{Enter}
/GO: Titles Second \title2{Enter}
/GO: Quit
 /G: Name Create Y_X{Enter}
 /G: View
```

The graph you have just made looks a lot like Figure 2.1 on page 35. However, it differs from the older graph in a number of important ways. In particular, your new graph does not reach as great a maximum height as did Figure 2.1. Also, it continues to curve downward for its entire length while Figure 2.1 is almost straight for times greater than about 8 seconds. Compare these two graphs and explain why they look so similar and yet differ in the ways noted. When you are finished, remember to resave DRAG2 with this new graph for future use!

EXERCISE 3.5^C Use the graph you created in Exercise 3.4 to explore the effect that various amounts of air drag have on the trajectory of the projectile. Values of the terminal velocity ranging from 7.5 m/s (considerable air drag) to 120 m/s (low drag) produce representative results. This numerical experiment is most easily done by positioning the cell pointer on cell C8, typing a new value for v_{term} followed by {Enter}, then pressing {F10} to view the resulting graph. When you have seen enough of this graph, pressing any key will bring you back to the worksheet, ready to repeat the process for a new value of v_{term}. After doing this for a number of values of the terminal velocity, print graphs showing the trajectory of the projectile for high, medium, and negligible air drag. Write a short paragraph stating which features of the trajectory change as air drag becomes more important. Support your conclusions with numerical results from the worksheet.

EXERCISE 3.6^C Make a graph that plots both the x-position and the height of the projectile versus time. You can do this in one of two ways. If you first clear the definition of the current graph with /Graph Reset Graph you can then define the new graph from scratch, using the prescription given in Exercise 3.4 as a guide. Alternately, you can start with Y_X as the current graph and modify it to make the new graph. Here the trick is to press {Escape} once whenever you are offered a default range or a default label that you wish to discard. Don't forget to change the labels on both the x-axis and the y-axis to reflect the variables you are now plotting.
    Regardless of the method you use to create this new graph, you will need to distinguish the two curves from one another. A prescription to do this is:

> /GO: Legend A x-position{Enter}
> /GO: Legend D y-position{Enter} Quit View
>   This assumes that the A-range is used for x and the D-range for y.

When you have finished defining this new graph, remember to give it a new name. A good choice would be XY_T. To include this graph in your disk file, save the worksheet DRAG1 again.
    Observe and explain the changes that occur in this graph as you vary the value of v_{term}.

EXERCISE 3.7^P Define a new graph, V_T, which plots the speed of the projectile as a function of time. Remember to label this new graph appropriately, create a new name for it, and resave the worksheet to preserve this newly defined graph. Observe how this graph changes as you vary v_{term}. Print representative graphs and comment on how they change with v_{term}.

3.5   ♠ Finding the Projectile's Range

When you launch a projectile it's important to know where it will land, as every sandlot ball player knows. The horizontal distance that the projectile lands away from its original position is called its range. As you may have already learned, when air drag is negligible the projectile has its maximum range on level ground for a given initial speed when launched at an angle of 45 degrees above the horizontal. `DRAG2` will let you determine the optimum launch angle when air drag is not negligible.

To begin this calculation, you should first modify `DRAG2` to accept the projectile's initial speed and launch angle rather than the Cartesian components, v_x and v_y, of its initial velocity. The following prescription will guide you through the changes needed:

B11: `/Worksheet Global Recalculation Manual`
B11: `/Worksheet Insert Row {Enter}`
B11: `"v_init`
C11: `50`
D11: `'m/s`
E11: `"th_init`
F11: `66`
G11: `'deg`
B11: `/Range Name Label Right {Enter}`
E11: `/Range Name Label Right {Enter}`
C12: `+$v_init*@cos($th_init*@pi/180)`
F12: `+$v_init*@sin($th_init*@pi/180)`
A19: `/Range Name Table {Enter}`
A19: `{GoTo} {Name} {Enter} {Calc}`

You can avoid displaying digits that are insignificant by choosing an appropriate format:

——: `/Worksheet Global Format Fixed 1{Enter}`
——: `/Range Format Fixed 2{Enter} dt{Enter}`

Although it is now easy to try different elevation angles (as the initial angle is named in ballistics), we can improve considerably the ease with which the achieved range can be recognized. One method is to look at a y-versus-x graph and find the x value where y returns to zero where it began. Another method is to scan down the columns of the work block to see where y falls back to zero and then record the corresponding x value on the notes page in a range column next to the entry for the elevation angle that produced that range. But a formula that reports the range on the notes page above the table would make the work go a lot faster. This is possible when we define the range as "maximum value of x while y is positive." (In all this discussion we have been assuming that $x_{init} = 0 = y_{init}$.) Spreadsheets allow a small amount of programming logic using an `@IF` function, and that is what you should use here.

——: `{GoTo}{Name}{Enter} {End}{Right} {Right}`
R37: `^range {Down}`

R38: ^[m] {Down}
R39: ^x_InAir {Down}
These are headings for a new work block column.

R40: @if(y>=0,x,"")
R40: /Copy R40 *to* R41..R140
R40: /Range Name Create X_INAIR{Enter} {Period}
{End}{Down} {Enter}
The column reports "if $y \geq 0$, x, otherwise nothing."

D16: ^ projectile range:
F16: @max(x_inair)
F16: /Range Name Create RANGE{Enter} {Enter}
G16: 'm
This produces on the notes page the desired report from the current computation.

The **syntax** (rules of usage—grammar and punctuation) for the @IF function are @IF(*condition,t_value,f_value*) where, as in all spreadsheet formulae, no spaces are allowed. The first argument of this function is a logical *condition*, a mathematical statement that will evaluate to true or false; in our application it is the statement y>=0, which is the way you must type $y \geq 0$. The second argument, *t_value*, is the value you want the cell to have when the condition is true; we let this be the value of x in the same row. The third argument *f_value* of @IF is the value you want the cell to have when the condition is false; we choose this to be a **string** (i.e., characters that write a label) containing no characters. If this null string seems too esoteric a value to suit you, try instead the string "underground" by entering @if(y>=0,x,"underground") in cell R40 and copying it down the column. The @MAX function that reports the range on the notes page regards all strings as having a numerical value of zero (but some spreadsheets may ignore strings in this context).

Searching for the Maximum Range

If you can remember the interesting cases (write them down on a note pad), you can search for the elevation angle that gives maximum range (for some particular initial speed) by simply changing th_init, recalculating the worksheet, and noting the range. In this way it should be easy to decide whether the maximum is to be found at, above, or below 45°. You would be well advised, however, to glance at the Y_X graph after each recalculation. (Use the command /Graph Name Use Y_X to make this graph current so it will be available with a press of {Graph}, i.e., {F10}, after each recalculation.) In some cases your choice for dt could be too small to compute the full range. Then @MAX(X_INAIR) may just report the last x-value you computed, not the true range out to $y = 0$. To automate this precaution, you can change the range formula in cell F16 to @IF(Y_LAST<=0,@MAX(X_INAIR),@ERR) after naming as

Y_LAST the last y-value computed (cell N140). Then the range formula will report an error if the projectile doesn't return to earth.

For a more systematic survey of the dependence of the range on the elevation angle th_init, lay out space for a range versus angle table on the notes page, which you then can graph after running a set of survey cases. The /Range Value command simplifies taking notes, as it copies the value of a cell (rather than the formula it contains) to another cell. This copied number, unlike a formula, will not change its value when the worksheet is recalculated later with different parameters.

> **G21: +th_init**
>
> **H21: +v_init**
>
> **G21: /Range Format Text {Right} {Enter}**
>
> > These formulae, which report the current values of the elevation angle and (after recalculation) the resultant range, can serve as column headings when their cells are formatted to show the formula text instead of its value.
>
> **G21: /Range Value {Right} {Enter} {Down} {Enter}**
>
> > This command copies the current values of θ_0 and the range to the row below.

To produce subsequent rows in the table, change the elevation angle, recalculate, and then use **/Range Value** to copy cells **G21** and **H21** to the first blank row in the table below them. A graph of this table may show unsightly jumps because our range formula picks the last x at which y was nonnegative, rather than estimating where x would be when, between our discrete time steps, $y = 0$. These jumps suggest a margin of computational error in our ranges that can be reduced only by more careful calculation.

EXERCISE 3.8[C]   Modify your worksheet to allow input of the elevation angle and output of the achieved range, as described in this section. Do a {PrintScreen} of the home screen after moving the cell pointer to cell F16 and pressing {Edit} so that the range formula is displayed.

EXERCISE 3.9[C]   Using $v_{term} = 30\,\mathrm{m/s}$ and $v_{init} = 50\,\mathrm{m/s}$, find the elevation angle that produces the maximum range. Print a graph or table showing several cases surrounding the maximum.

3.6   Your New Spreadsheet Skills in Review

- Block structure to simplify editing.

- Using {GoTo}{Name}, i.e., {F5}{F3}, to jump to named cells with # the first name on the menu, allowing easy access to the work block.

- Using the {Edit} key, {F2}, to allow changes in long formulae without retyping the useful parts.

- Moving blocks of working formulae with **/Move**.

- Deleting rows or columns with **/Worksheet Delete Row** or **/Worksheet Delete Column**.

- Inserting rows or columns with **/Worksheet Insert Row** or **/Worksheet Insert Column**.

- Taking notes with the **/Range Value** command.

- More flexible control of printing than {PrintScreen} allows by accessing the **Print** menu:

    - Select a range (block) for printing with **/Print Printer Range**
    - Begin printing with **/PP Go** after making any other choices in the **/Print Printer** menu.
    - To list the underlying formulae instead of the data that appear on the screen, use **/PP Options Other Cell-formulas**.

3.7   Listing

```
range names:
TITLE2  A1      #    J36
TITLE1  D2      T    J39
G       C7      X    K39
V_TERM  C8      VX   L39
X_INIT  C10     AX   M39
Y_INIT  F10     Y    N39
VX_INIT C11     VY   039
VY_INIT F11     AY   P39
DT      C13     |V|  Q39
```

```
J38: ^t                     K38: ^x                   L38: ^vx
J39: 0                      K39: +X_INIT              L39: +VX_INIT
J40: +T+$DT                 K40: +X+VX*$DT            L40: +VX+AX*$DT
J41: +J40+$DT               K41: +K40+L40*$DT         L41: +L40+M40*$DT
...                         ...                       ...
J139                        K139                      L139


M38: ^ax                    N38: ^y                   038: ^vy
M39: -$G*VX*|V|/$V_TERM^2    N39: +Y_INIT              039: +VY_INIT
M40: -$G*L40*Q40/$V_TERM^2   N40: +Y+VY*$DT            040: +VY+AY*$DT
M41: -$G*L41*Q41/$V_TERM^2   N41: +N40+040*$DT         041: +040+P40*$DT
...                         ...                       ...
M139                        N139                      0139


P38: ^ay                    Q38: ^|v|
P39: -$G-$G*VY*|V|/$V_TERM^2 Q39: (VX*VX+VY*VY)^0.5
P40: -$G-$G*040*Q40/$V_TERM^2 Q40: (L40*L40+040*040)^0.5
P41: -$G-$G*041*Q41/$V_TERM^2 Q41: (L41*L41+041*041)^0.5
...                         ...
P139                        Q139
```

Listing 3.1: The DRAG2 worksheet.

Chapter 4

Numerical Efficiency

Our approach to mechanics has one dominant theme: Understand the causes of accelerations, which are called forces, and then use a predictable acceleration a to give "marching orders"

$$
\begin{aligned}
dv &= a\,dt \\
dx &= v\,dt
\end{aligned}
\tag{4.1}
$$

to update successive changes in a particle's velocity v and position x. In nature as envisioned by classical mechanics these changes are occurring on arbitrarily small (infinitesimal) time scales dt, but in computer models we use finite (nonzero) time steps dt. This chapter asks when dt is small enough, and draws a scheme from the branch of mathematics known as numerical analysis that dramatically improves our ability to get by with moderately large dt's.

Because these marching orders build a graph of $x(t)$ only when all the small changes given by the equations in 4.1 are added up, the process is called integration. We have previously used the Euler integration method, which rewrites equations 4.1 as

$$
\begin{aligned}
v_{n+1} &= v_n + a_n\,dt \\
x_{n+1} &= x_n + v_n\,dt \quad .
\end{aligned}
\tag{4.2}
$$

Here x_n, v_n, and a_n are the values of x, v, and a at time $t_n = t_0 + n\,dt$ that appear in the nth row after the beginning of the work block computations. Similarly, x_{n+1}, ..., are those in the following row. In the next section we present an example that shows how difficult it is to obtain good accuracy by this method.

The numerical integration method we recommend (in cases where a does not depend explicitly on the velocity) is not very different from this and requires the same number of computations. It is called the leapfrog or half-step method and treats the velocity (in a given row of the work block) as belonging to a

time a half step later than the position and acceleration in that same row. The
mathematical specification for it is

$$
\begin{aligned}
x_{n+1} &= x_n + v_{n+1/2}\,dt \\
v_{n+3/2} &= v_{n+1/2} + a_{n+1}\,dt \quad .
\end{aligned}
\tag{4.3}
$$

The rationale is that, for finite dx and dt, the ratio $v = dx/dt$ is an average
velocity over the interval dt and is more reasonably associated with the center
of the interval than with one end. Similarly, for finite differences dv and dt,
the ratio $a = dv/dt$ should be associated with the midpoint time between the
two v values in $dv = v_{n+3/2} - v_{n+1/2} = a_{n+1}\,dt$. This slight change often lets
us compute in 100 rows to an accuracy that would have needed thousands of
worksheet rows by the Euler method.

4.1   ITEST: Testing Integration Schemes

We will build a worksheet incorporating the leapfrog method, but first we'll
solve a well understood problem by the Euler method. Minor changes then
convert it to the leapfrog method, and the contrasts are illuminating.

The limitations of the Euler method do not appear in problems with constant acceleration where standard texts teach how to describe the motion analytically, but are prevalent in cases of nonconstant acceleration where computers are usually required. One particular case of nonconstant accelerations—the
situation where the acceleration a is proportional to the displacement x—is the
physicist's favorite both because it has an analytic solution and because it arises
as a useful approximation in many realistic applications. (You will meet it as
physics in later chapters on oscillations; here we treat it as mathematics.) We
choose this as our test case and apply our spreadsheet skills to solve $a = -x$.

What do we mean: "Solve $a = -x$"? (Hint: you solved $a = -g - gv^2/v_{\text{term}}^2$
in Chapter 2 and a similar problem in Chapter 3.) By **solving** $a = -x$ we mean
treating a as a derivative d^2x/dt^2, translating $d^2x/dt^2 = -x$ into the marching
orders of equations 4.1 (with $a = -x$) and, via the spreadsheet, obtaining
graphs of $x(t)$. Mathematicians call $d^2x/dt^2 = -x$ a differential equation, and
treat them in advanced courses in calculus. But with the aid of the computer,
we can (as in previous chapters) deduce its consequences without any more
difficulty than would be required to make the spreadsheet describe a savings
account accumulating compound interest (for which see Exercise 4.6).

This approach can also be stated in physics terms: The equation $a = -x$
can arise from Newton's second law $a = F/m$ with a force law $F/m = -x$
(which occurs for a mass attached to a suitable spring that pulls back with
increasing force as it is stretched larger distances x beyond its natural length).
Here we have a theory that predicts the acceleration of a mass when its position
is known, $a = -x$. This information is then used to update successively the

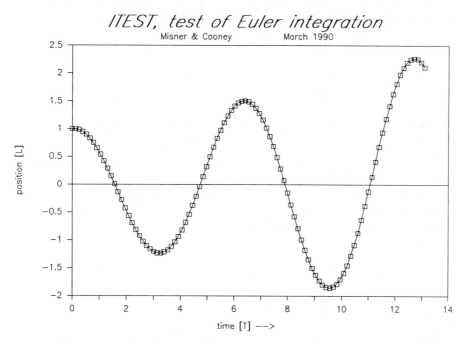

Figure 4.1: A motion with $a = -x$ solved using the Euler method.

velocity and position of the moving mass from equations 4.1, yielding a graph of its motion.

Since you have done this before, you can retrieve and edit a previous worksheet, such as the worksheet from Section 2.4, for which we suggested the filename **DRAG1**. The result is shown in Schema 4.1, Screen 4.1, and Listing 4.3, where appropriate labels identify this problem and its parameters. Exercise 4.1 (page 62) asks you to build this worksheet, named **ITEST**, which you should do now. You will save some time if you issue the **Worksheet Global Recalculation Manual** command (cf. page 19) soon after starting to work.

If you produce the **ITEST** worksheet by editing **DRAG1**, you will find the data still being graphed properly and most of the graph titles still appropriate. For instance, when you change the heading in cell **D2** to read **ITEST, test of Euler**...(as suggested in Screen 4.1), the main title of your graphs also changes since this cell (named **title1**) was assigned as the title for all graphs when you wrote the **DRAG1** worksheet. A few references to **height** instead of **position**, however, do need changing (Exercise 4.2).

Checking the Solution Your worksheet should produce a graph like Figure 4.1, for instance, and a graph **V_T** of velocity versus time (Exercise 4.3). If you used the same initial conditions as these examples, $x_0 = 1$ and $v_0 = 0$, then

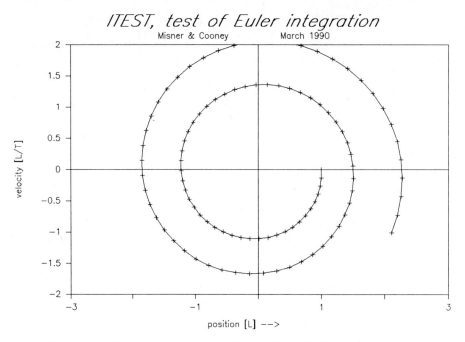

Figure 4.2: Phase-plane plot of motion with $a = -x$ solved using the Euler method. To obtain an aspect ratio similar to this example, you must use commands from the /Graph Options Scale menu to set the limits on v and x manually; automatic scaling often makes circles look like ellipses.

the exact mathematical solution to $d^2x/dt^2 = -x$ is

$$\begin{aligned} x &= \cos t \\ v &= -\sin t \end{aligned} \tag{4.4}$$

as you will verify at some point in your calculus course.[1] You should see substantial similarities between your graphs X_T and V_T and these formulae for x and v, but also important differences that are signs of the inaccuracies in the numerical approximations used in the ITEST worksheet. If you have forgotten exactly what a graph of $\cos t$ looks like, or want to compare the numerical (Euler) solution with the exact one in detail, add a column x_exact to the work block, tabulate $\cos t$ there, and plot it on your X_T graph as range D so that you can see both the Euler approximation and the exact solution on the same graph (exercise 4.4). The success of the Euler method is that it

---

[1] You can verify this analytic solution only if you have had more than a couple months of calculus. One verifies the $x = \cos t$ solution of $d^2x/dt^2 = -x$ by first calculating from it $v \equiv dx/dt = (d/dt)\cos t = -\sin t$ and then $a \equiv dv/dt = (d/dt)(-\sin t) = -\cos t = -x$, i.e., $a = -x$. If these computations are clear to you, you should also be able to show analytically that $x = \sin t$ is another solution, as is any linear combination of $\sin t$ and $\cos t$.

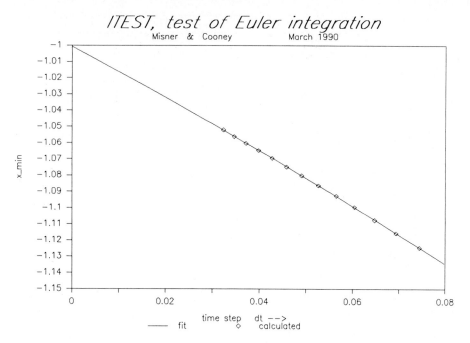

Figure 4.3: The effects of change in dt on the calculated value of the first minimum of x based on $x_{init} = 1$, $v_{init} = 0$, as in Screen 4.1 (page 63). The smooth curve is a quadratic polynomial in dt fit to the calculated points to provide an extrapolation to $dt = 0$ where the analytic solution would give $x_{min} = -1$.

correctly shows that $x(t)$ oscillates about $x = 0$, and it even gets the period of oscillation (time interval between maxima of x) about right. But it fails badly to produce an obvious property of the exact $\cos t$ graph, namely that all the maxima are equal. The source of this error is the computer's requirement that dt to be finite (nonzero) in equations 4.1 while the analytic solution solves these equations in the limit $dt \to 0$, i.e., with infinitesimal dt.

Another view of the approximation errors the Euler method produces is given in Figure 4.2. This is called a phase-plane plot, meaning a display of the motion in the vx plane (or often the px plane where p is momentum). For the exact solution given by equations 4.4, this graph should be a circle, since the trigonometric identity $\sin^2 t + \cos^2 t = 1$ gives, from equations 4.4, $v^2 + x^2 = 1$, which is the equation of a circle in the vx plane. Numerical error in the Euler approximation has converted this circle into a spiral in Figure 4.2.

Will the Euler method give a good approximation to the limit if we choose dt small enough? To test this, run the same calculation several times, using different time steps for dt to see how the results change. Of course, the graphs change dramatically (as you change dt) because, for the same number of time steps (rows in your worksheet), a different range of physical time ($t_0 = 0$ to

labels:	t	x	v	a

init data:	t_0:
	      0
		x_0:
		      x_{init}
			v_0:
			      v_{init}
				a_0:
				 $F(x_0)/\mathbf{m}$

typical line:	t_1: $t_0 + \mathbf{dt}$	x_1: $x_0 + v_0\,\mathbf{dt}$	v_1: $v_0 + a_0\,\mathbf{dt}$	a_1: $F(x_1)/\mathbf{m}$

Schema 4.1: Euler integration scheme for any force $F(x)$. For the ITEST worksheet, use $-x$ for F/m.

$t_{100} = 100\,dt$, say) is being defined. Figure 4.3 picks out a fixed physical phenomenon, the first minimum in position after starting with the initial conditions of Screen 4.1 and plots it against dt. To make a graph like this, simply type out the results of a few cases as a table somewhere on your notes page. To run each case, change dt and look down the position column for the first minimum. More simply, read the minimum x value from the I/O page and record it beside dt in columns on the notes page, but only in cases where it is the first minimum. That you check by looking at a graph to see that the first minimum, which you want, is the absolute minimum, which the formula @MIN(L44..L144) in cell F11 reports. (Exercise 4.7 shows how this tabulation can be partly automated. Use of the /Data Table 1 command can automate it further.)

No Quick Fix   Figure 4.3 indicates that our approximation improves when dt is small. But using a very small dt, such as $dt = \pi/1000$ is impractical. Although that choice gives the first minimum accurate to 0.5%, the tenth minimum is still in error by about 10%. And it would require 2000 rows just to plot one cycle of oscillation. Your computer could run out of memory, and, even if it did not, recalculations and graphing times would be very slow. The next section gives a better solution.

EXERCISE 4.1[C]   Build the worksheet ITEST described in Schema 4.1, Screen 4.1, and Listing 4.3. If you have previously built worksheet DRAG1, its blocked form from Exercise 3.1 (page 46) provides an excellent starting point. Just be sure to save it under its new name, ITEST, before you begin modifying it—and again frequently as you make progress. Use /Range Erase to clear away areas of the home screen that can't be reused before retyping the changed text and revising the acceleration formula. You should also clear away the unneeded range names H_INIT, H, G, and V_TERM (using the /Range Name Delete command) and install appropriate new ones:

     C10: /Range Name Create X_INIT{Enter} {Enter}
     --: {GoTo} #{Enter}
     I36: /Worksheet Insert Row I36..I40
         After assigning names on the home screen, move to the work block and insert rows as needed if you want your file to be organized exactly as our sample was in Screen 4.1.
     L43: ^x

C13: U @PI/24   **READY**

```
         A          B          C          D          E          F          G          H
 1  your name  &  today's date
 2  your class & section           ITEST, test of Euler integration
 3
 4
 5  --------------------------------------------------------------------------
 6  Construction parameters:
 7          None.   Mathematical example:
 8                        a = -x
 9  Initial data:                    Results:
10          x_init =        1        x_max = 2.275606
11          v_init =        0        x_min = -1.85365
12  Approximation:
13             dt = 0.130899
14  --------------------------------------------------------------------------
15  Notes page:
```

M45: +V+A*$DT   **EDIT**
+M44+N44*$C$13

```
         I          J          K          L          M          N          O          P
41  name:                   time       pos'n      vel        accel
42  units:                  [T]        [L]        [L/T]      [L/T^2]
43  labels:                 t          x          v          a
44  init data:              0          1          0          -1
45  typical row:            0.130899   1          -0.13089   -1
46  copied rows:            0.261799   0.982865   -0.26179   -0.98286
47                          0.392699   0.948595   -0.39045   -0.94859
```

Screen 4.1: Worksheet layout for ITEST which provides an Euler method integration of $d^2x/dt^2 = -x$.

K43: /Range Name Labels Down K43..N43
>   Update the work block labels and assign them to the first computational cells.

In the report block on the I/O screen list the maximum and minimum x values found in the work block:

F10: @max(L44..L144)
F11: @min(L44..L144)

As deliverable output of this exercise, produce a {PrintScreen} copy (like Screen 4.1) of your screen layout and a listing (see Section 3.3 on page 48) of your principal formulae (corresponding to rows 43..46 in Screen 4.1).

EXERCISE 4.2[C]   Change **height** to **position** and H to X as needed in the graphs in the ITEST worksheet. Remember to use the /Graph Name Create command to store your changes (and remember, as always, to save the worksheet after you have modified it). To change the names of the graphs, e.g., H_T to X_T, you must first make current a copy of the old graph (/Graph Name Use H_T), modify that current graph to make the graph you want, preserve it by using /Graph Name Create X_T with the desired name, and then delete the old graph (/Graph Name Delete H_T), which your changes have not modified. Print the X_T graph.

EXERCISE 4.3$^\text{C}$ Design and print graphs V_T, and V_X (i.e., v versus t and v versus x) from the ITEST worksheet. The A_V graph from DRAG1 is not significant (because a is not given as a function of v now) and can be deleted.

EXERCISE 4.4$^\text{E}$ Add a column to the work block of the DRAG1 worksheet that tabulates the function $x_{\text{exact}} = x_0 \cos t + v_0 \sin t$ and plot it on the same X_T graph as the numerically integrated $x(t)$ solution. (Since this is a mathematical exercise, we are using a dimensionless time. This allows x_0 and v_0 also to be treated as dimensionless numbers.) Use the /Graph Options Format command to plot the Euler $x(t)$ as Symbols only and the exact $x(t)$ as Lines so that they are clearly distinguished on a black-and-white printout. Identify each quantity, using /Graph Options Legend. Print such a graph.

EXERCISE 4.5$^\text{E}$ Modify the ITEST worksheet to use the force law $a = F/m = -1$. Be sure to change the title in cell D2 before printing graphs. Print graphs X_T and V_T; relate the motions they show to an analytic solution of Newton's laws for this force and your chosen initial conditions.

EXERCISE 4.6$^\text{P}$ Build a worksheet that calculates how a given amount P_0, the initial principal, grows in a savings account as it accumulates compound interest. Print your home screen showing the I/O area and provide a listing of the first few rows of your work block. If I is the annual interest rate and dt the compounding interval (in years or fractions), then the new balance in the account after the interest for the nth interval has been paid, will be $P_n = P_{n-1} + I\,dt\,P_{n-1}$. Note that I must be stated as a fraction, e.g., 0.085 for 8.5%, but the cell displaying it in the I/O area can be formatted to display it as a percentage.

EXERCISE 4.7$^\text{E}$ Produce a graph like Figure 4.3 (but less detailed) showing how the first minimum of $x(t)$ depends on the interval size dt used in the approximation. To make note taking convenient, first make a table heading using formulas for the data you want:

        D19: +DT
        E19: +X_MIN
        E19: /Range Format Text E19..D19

Then position the screen so that both the dt input cell and the range you will use for the table (below D19 and E19) are visible. Each time you change dt you should check a graph (using {Graph}, the F10 key), previously selected as either X_T or V_X, to verify that the minimum x is the first minimum. When you have a case that you want to record, use the /Range Value command to copy D19..E19 to a blank line in the table below it. When graphing this table, use /Graph Options Format Graph Symbols to prevent lines crisscrossing the graph if the data are out of order. Use manual scaling from the /Graph Options Scale menus to make sure $dt = 0$ and $x_{\text{min}} = 0$ are visible on the graph, so your eye can extrapolate the data there. (Our graph shows also a fitted line from data tabulated by an advanced feature of the spreadsheet—the /Data Regression commands.)

labels:	t	x	vhs	a
init data:	t_0: 0	x_0: x_init	$v_{1/2}$: $v_\text{init} + a_0\,dt/2$	a_0: $F(x_0)/\text{m}$
typical line:	t_1: $t_0 + \text{dt}$	x_1: $x_0 + v_{1/2}\,\text{dt}$	$v_{3/2}$: $v_{1/2} + a_1\,\text{dt}$	a_1: $F(x_1)/\text{m}$

Schema 4.2: Leapfrog integration scheme for any force $F(x)$. In the **LEAP** worksheet, set $F/m = -x$.

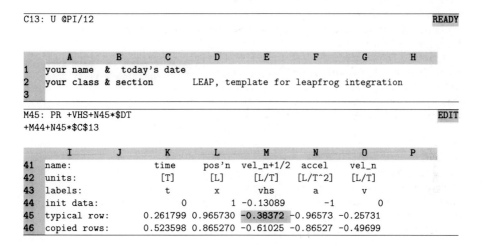

Screen 4.2: The **LEAP** worksheet.

4.2 LEAP: Leapfrog Integration

Few changes in the **ITEST** worksheet are needed to convert it to the leapfrog algorithm of equations 4.3. Schema 4.2 shows these equations in a spreadsheet arrangement and differs from the Euler algorithm in Schema 4.1 (page 62) only in the velocity column which is now labeled **vhs** to suggest velocity at the half step. Load the previous Euler version **ITEST** if you have been away from the computer since completing it. Press {Calc}, the F9 key, to be sure the results are valid; then use {Graph} to recall the solution the Euler method gave. We will now make the most important change to see the difference it makes.

> **M45:** {Edit} {Insert}
> **M45:** +M44+N45*$DT
> **M45:** /Copy M45 *to* M46..M144
>> Edit this typical velocity-update cell in overtype mode (which {Insert} toggles) to change the acceleration used previously (cell **N44** in the row above) to the acceleration in the same row (cell **N45**). Then copy this change down the column.

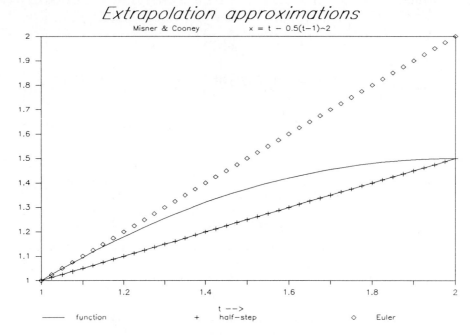

Figure 4.4: Comparison of the Euler and half-step extrapolations for a quadratic polynomial.

−−: {Calc} {Graph}

The wild swings of increasing amplitude that concerned us in Figure 4.1 should now be replaced by repetitious, or periodic, motion where the displacement does not increase from one cycle to the next. Far from needing a much smaller time step dt to keep the amplitude error under control, we can now use a larger time step: try $\pi/12$ or $\pi/8$ instead of the $\pi/24$ that yielded Figure 4.1. The lack of huge amplitude errors does not, of course, mean the results are correct. To be convinced that some characteristic you see in the numerical approximation is actually predicted by the differential equation, you must verify that it remains essentially unchanged when dt is decreased (and the number of rows computed is correspondingly increased as needed).

Rename the modified worksheet to **LEAP.wk1** by saving it under the new name **LEAP**. Save it again after you make each of the further improvements in Exercises 4.8 and 4.9 to complete this leapfrog template as shown in Screen 4.2. This requires changing the $v_{1/2}$ formula as well as a few labels and cell names. The additional column **O** you see in that screen was not needed to solve the differential equations 4.1 but is needed for the phase-plane graph **V_X** as illustrated in Exercise 4.9.

EXERCISE 4.8[C]   Modify the **ITEST** worksheet to use the leapfrog method, as described earlier in this section, and rename it **LEAP**. In Screen 4.2, note the label

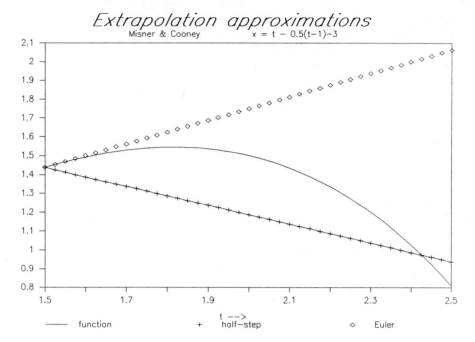

Figure 4.5: Comparison of the Euler and half-step extrapolations for a cubic polynomial.

changes in cells D2, M41, and M43; in Schema 4.2 note the formula modifications needed in the first and subsequent cells (for the half-step velocity) in column M. Also delete the table and x_{min} versus dt graph associated with the Euler data from ITEST.

EXERCISE 4.9[C] Using the large step size $dt = \pi/8$, view the V_X graph that plots the column M velocity against the column L position. It appears to show what calculus teaches is impossible: the minimum x occurs at a nonzero v. But these values refer to different times; the plotted points are $(x_n, v_{n+1/2})$. We need a graph plotting points (x_n, v_n). Complete the LEAP worksheet by adding column O for $v_n = (v_{n-1/2} + v_{n+1/2})/2$ (as specified in Listing 4.1) and changing graphs V_X and V_T to use this column for v.

4.3 ♠ Numerical Analysis

This section amplifies the introductory remarks on page 58 following equations 4.3 about the mathematical basis for the leapfrog method. (For a higher level treatment, see an article by Ronald Ruth in *IEEE Transactions*, NS vol. 30, no. 4, Aug. 1983, p. 2669.)

One Time-Step in Detail Figures 4.4 and 4.5 each graphically compare two extrapolations using derivatives. These graphs are to be viewed as examinations of a single time step under a microscope; the two ends of the x-axis represent the beginning and end of just one step, dt, in the evolution of some motion, $x(t)$. Both the Euler and half-step methods use a straight line to extrapolate from the beginning value, $x(t)$, to the end value, $x(t + dt)$; the difference is in the ways they assign the slope to the straight line. Suppose that $x(t)$ is the exact solution we are trying to find, and $v(t) = dx/dt$ its derivative. The Euler method, which we have used since Chapter 2, is based on the estimate $x(t + dt) = x(t) + v(t) \, dt$, which uses the slope at the start of the interval to make a linear extrapolation. This is not exact. In calculus, the mean-value theorem states a similar but exact relationship—

$$x(t + dt) = x(t) + v(t + \xi dt) \, dt \quad , \quad 0 < \xi < 1 \tag{4.5}$$

—however, the value of ξ is not predicted specifically by the theorem. It turns out that $\xi = 1/2$, which uses the slope at the middle of the interval, is usually a better guess than $\xi = 0$. In fact, using the worksheet APROX, defined in the listings at the end of the chapter, we can do numerical experiments and find that, for quadratic functions $x(t) = x_0 + v_0 t + at^2/2$, the half-step estimate is exact, which is also easily proven analytically. For more general functions $x(t)$, it turns out that halving the step size dt in the Euler method halves the error, as seen in Figure 4.3. With the leapfrog method, however, halving the step size cuts the error by a factor of four.

4.4   Listings

```
I44: 'init data:   K44: 0        L44: +X_INIT
I45: 'typical row: K45: +T+$DT   L45: +X+VHS*$DT
I46: 'copied rows: K46: +K45+$DT L46: +L45+M45*$DT   K47..O144:  ...

                   M44: +V+A*DT/2      N44: -X      O44: +V_INIT
                   M45: +VHS+N45*$DT   N45: -L45    O45: (M45+VHS)/2
                   M46: +M45+N46*$DT   N46: -L46    O46: (M46+M45)/2

range names:
X_INIT C10   TITLE1 D2   #   I41
V_INIT C11   TITLE2 A1   T   K44
DT     C13               X   L44              A15: 'Notes page:
                         VHS M44
                         A   N44              A17: U 'range names:
                         V   O44
```

Listing 4.1: The LEAP worksheet.

```
          Graph Settings                           Graph Settings
Name:    V_T            Type: XY         Name:    V_X            Type: XY
Range                   Format           Range                   Format
   X:    K44..K144                          X:    L44..L144
   B:    O44..O144   Lines                   B:    O44..O144   Both
```

Listing 4.2: Graph revisions for **LEAP**.

```
A12: U 'Approximation:
B13: "dt =        C13: U @PI/24
E9:  U 'Results:
E10: "x_max =    F10: @IF(L144>0#AND#M144>0,@NA,@MAX(L44..L144))
E11: ^x_min =    F11: @IF(L144<0#AND#M144<0,@NA,@MIN(L44..L144))

I43: 'labels:        K43: ^t        L43: ^x
I44: 'init data:     K44: 0         L44: +X_INIT
I45: 'typical row:   K45: +T+$DT    L45: +X+V*$DT
I46: 'copied rows:   K46: +K45+$DT  L46: +L45+M45*$DT
                     K47..L144:  ...

range names:                           M43: ^v            N43: ^a
X_INIT  C10    TITLE1  D2    #  I41     M44: +V_INIT       N44: -X
V_INIT  C11    TITLE2  A1    T  K44     M45: +V+A*$DT      N45: -L45
DT      C13    X_MAX   F10   X  L44     M46: +M45+N45*$DT  N46: -L46
               X_MIN   F11   V  M44     M47..N144:    ...
                             A  N44
```

```
          Graph Settings                           Graph Settings
Name:    V_T            Type: XY         Name:    V_X            Type: XY
Titles:                                Titles:
X-axis:  time [T] -->                  X-axis:  position [L]  -->
Y-axis:  velocity [L/T]                Y-axis:  velocity [L/T]

Range                   Format         Range                   Format
   X:    K44..K144                        X:    L44..L144
   B:    M44..M144   Lines                 B:    M44..M144   Both
```

Listing 4.3: **ITEST**: Euler method integration of motion with $a = -x$. As
in Listing 2.1, some cells have decorative emphasis indicated by a U prefix which is provided by the **Range Unprotect** command. The formulae
in cells **F10** and **F11** have been elaborated beyond the text prescriptions
to yield a warning **NA** in some cases where the result would not be meaningful. Use {Help} when you want to learn about the **@IF** function and
the **#AND#** logical operator.

```
B6: [W11] "dt          C6: U [W7] 1/40
B7: [W11] "midslope     C7: [W7] ($C33-$C31)/2/$DT
B8: [W11] "startslope   C8: [W7] ($C13-X)/2/$DT

          E6: "t_init              F6: U 1
          E7:
          E8: 'x = t - 0.5(t-1)^2

B10: [W11] ^t           C10: [W7] ^x
B11: [W11] +T0-$DT      C11: [W7] +T-0.5*(T-1)^2
B12: [W11] +T_INIT      C12: [W7] +T0-0.5*(T0-1)^2
B13: [W11] +T0+$DT      C13: [W7] +B13-0.5*(B13-1)^2
B..: [W11]  ...         C..: [W7]  ...
B52: [W11] +B51+$DT     C52: [W7] +B52-0.5*(B52-1)^2

          D10: ^Euler                 E10: ^half-step
          D11:                        E11:
          D12: (T0-$T0)*$STARTSLOPE+$X0   E12: (T0-$T0)*$MIDSLOPE+$X0
          D13: (B13-$T0)*$STARTSLOPE+$X0  E13: (B13-$T0)*$MIDSLOPE+$X0
          D..:  ...                   E..:  ...
          D52: (B52-$T0)*$STARTSLOPE+$X0  E52: (B52-$T0)*$MIDSLOPE+$X0

range names:                                I3: U 'title2:
DT        C6    T_INIT F6    TITLE1 D2       J3: +A1&"              "&E8
MIDSLOPE  C7                 TITLE2 J3
STARTSLOPE C8
T         B11 X     C11
T0        B12 X0    C12
```

Graph Settings

Name:	COMPARE		Type:	XY	
Titles:		Range		Format	Legend
X-axis:	t-->	X:	B10..B110		
1st:	\title1	A:	C10..C110	Lines	function
2nd:	\title2	B:	E10..E110	Both	half-step
		C:	D10..D110	Symbols	Euler

Listing 4.4: Figure 4.4 is produced from this APROX worksheet. For Figure 4.5 the exponents ^2 must be changed to ^3 and a different t_{init} selected.

Chapter 5

Force Causes a Change in Velocity

This chapter's title states the principle at the heart of Newtonian physics. This principle is also the effective method for building spreadsheet models of mechanical systems in motion.

5.1   Newton's Second Law

Physicists like to find a logical path that solves specific problems from first principles; here that path is very short. In previous chapters, you have seen that when a projectile's acceleration is specified, most aspects of its motion are determined, as in the examples of projectiles influenced by gravity and air resistance. Newton's second law gives the basic outline for this understanding of motion:

$$\mathbf{a} = \mathbf{F}/m \quad . \tag{5.1}$$

In words, this says that acceleration is caused by force and will have a value $\mathbf{F}/m$. By using a spreadsheet, then, whenever we can supply a formula $\mathbf{F}(\mathbf{r}, \mathbf{v}, t)$ for the force acting on a particle, we can proceed to calculate in a succession of small time steps, dt, the particle's successive changes in velocity and position:

$$d\mathbf{v} = (\mathbf{F}/m)\,dt \qquad \text{and} \qquad d\mathbf{r} = \mathbf{v}\,dt \quad . \tag{5.2}$$

Accurate numerical methods to approximate these formulae, which are intended to apply exactly only in the limit $dt \to 0$, are simplest in the case where the force is velocity independent, as seen in the previous chapter. We will begin with that case.

Single Particle Mechanics   At first, we consider only a very simple mechanical system, namely a single particle moving in one dimension. Simple as it is, this system is the paradigm for a Newtonian understanding of motion. Physicists develop much of their intuition for mechanics by studying one-dimensional, single-particle problems. You will learn later how more complicated and interesting problems can sometimes be reduced to this case. This chapter contains some hints about how the examples explored here relate to those other problems.

Framework and Examples   The Newtonian viewpoint expressed in the title of this chapter is a mindset or mental framework physicists often adopt. This viewpoint is the most important thing you can learn this semester; to help you exercise it, we will provide a variety of examples in which it can be applied.

   To prevent your thinking that friction or mg or some other simple example of a force is a central point of our study now, we will introduce several varieties of force at once with only sketchy suggestions of where we think they are applicable. This procedure allows you to focus on the wider framework: any specification of a rule to determine the total force on a particle will lead to a detailed description of that particle's possible motions. Once this framework has become a working tool you can question the different force laws and ask which are important in the real world. The answer to this question (which, in a wider context, still remains at the forefront of physics research) deems gravitational and electromagnetic forces as more fundamental than others at the everyday level of experience.

Force Prototypes

Four different force laws are considered here; you should try using at least three of them. They are:

$$F_1 = -1 \tag{5.3}$$

$$F_2 = -x \tag{5.4}$$

$$F_3 = \frac{1}{2}\left(\frac{1}{x^5} - \frac{1}{x^3}\right) \tag{5.5}$$

$$F_4 = \frac{x}{2}(1 - x^2)\left[\frac{5}{2 + 3x^2}\right]^3 . \tag{5.6}$$

You see no adjustable parameters in these formulae, not even a gravitational acceleration g or a mass m in F_1. These formulae are simple enough that all adjustments can be absorbed into the choice of units (almost all—the numerical coefficients in $[5/(2 + 3x^2)]$ from equation 5.6 could be different). These formulae should not be interpreted as using SI units. Instead, the units of length, mass, and time are anything you want them to be. For instance, to treat the motion of a one-stone mass thrown up in the air, let $M = 1$ U.K. stone $= 14$ lb-mass be the unit of mass and $L = 9.8$ m be the unit of length. The gravitational force on a one-stone mass m would then be $F_1 = -mg = -(1\,M)(1\,L/s^2)$, which can be entered into the computer simply as $F_1 = -1$. Thus, F_1 can represent any constant force (by appropriate definition of mass and length units). Similarly, F_2 can represent any linear restoring force $F_2 = -kx$ by appropriately defining the units of mass and time.

Textbook Forces   Motions controlled by forces F_1 or F_2 can be solved analytically (although the F_2 case usually is covered in a later chapter, on simple harmonic motion, so we will not introduce all the vocabulary to describe that motion here). Net, or total, forces in the F_1 class impart constant accelerations to the masses they act on; they are the subject of most of the paper-and-pencil exercises done at this point in the course. But forces F_3 and F_4 control motions that cannot be solved in detail analytically, so you must write computer worksheets to solve for these motions, as we will explain.

Chemical Bonding Force   The F_3 force is a simplified model for the chemical bonding force that acts between, say, two oxygen atoms in an O_2 molecule. Here we imagine that we look at the motion of only one of the atoms, so x represents its location while its partner dances mirror-image steps at $-x$. Notice from the formula defining it, that the force is zero at $x = 1$. Consequently, the particle we follow can sit quietly at that location without any force acting to move it away. This implies that the unit of length L (implicit in the use of this formula for force) is related to the length of the chemical bond we try to model: the normal distance between the two atoms here has the value $2 L$. A worksheet in Appendix D graphs F_3 as a function of x (see Screen D.2, page 215). From such graphs, or directly from the formula for F_3, you can see that F_3 is positive/repulsive (pushes in the positive x-direction) when $0 < x < 1$ and is negative/attractive when $x > 1$. To make these signs more explicit, we can rewrite the formula as

$$F_3 = \frac{1}{2x^5}\left(1 - x^2\right) \quad . \tag{5.7}$$

The graph (or the formula) also shows that F_3 gets arbitrarily large for x near zero and arbitrarily small for very large x.

Force Between Magnets F_4 represents the force exerted (by some fixed center of force at $x = 0$) on a moving particle. It will, like F_3, reappear when we progress to the study of interactions between two particles and of collisions with conservation of energy and momentum. It should be not very different from the force a fixed pair of parallel bar magnets would exert on a similar magnet that could move only along a third parallel line between them on a frictionless air track. (See Figure 8.2, page 121.) Notice in this case the three positions of zero force: $x = 0$ and $x = \pm 1$, corresponding to the movable magnet exactly between the fixed magnets or at a distance of ± 1 unit away.

5.2   CBOND: **Molecular Vibration**

Let us begin by studying the motions allowed by the force F_3, proposed as similar to the forces binding atoms together in molecules. From Newton's second law, $a = F/m$, and from the law of force F_3, in equation 5.5, we have a theory that predicts the acceleration of the mass (one atom in the molecule) when its position is known. We can then use the efficient techniques of Chapter 4 to add up the successive small changes in velocity and position implied by this acceleration and plot the motions. All that is needed is to change the formula for the acceleration in the template (model worksheet) LEAP and explore the motions it computes. For good housekeeping, the I/O page should be edited to reflect the problem at hand, and the worksheet should be renamed, e.g., CBOND. This is specified in Exercise 5.1. Screen 5.1 on page 77 shows portions of CBOND.

Explore

Use $dt = \pi/25$ and find several qualitatively different motions by trying both large and small values of x_{init} and v_{init}. Note any cases where the choice of dt might need review. Look at all three graphs (X_T, V_T, and V_X) for any case that looks interesting or typical and try to visualize the actual motion and see how the information in the X_T graph is being seen merely from different viewpoints in the other graphs. At this point, think of the mathematical model as a new toy. Push it, pull it, wind it up various amounts, then turn it loose and see how it behaves.

Each set of initial conditions (x_{init}, v_{init}) should fit into one of two categories. The first category is **bound state**, where $x(t)$ will for all t in the interval $(-\infty, +\infty)$ remain finite $|x(t)| < x_{large}$. The other is **scattering state** where $|x(t)| \to \infty$ as $t \to \pm\infty$.

The proposed interpretation of this model is that $x(t)$ is the position of one atom in a diatomic molecule, such as O_2 or N_2. A bound state is one where the two atoms stay near each other forever; a scattering state is where they may briefly be close but start and end far apart. The bound state corresponds to the normal molecule, perhaps somewhat jostled; the scattering state corresponds to atoms in a hot, rarefied gas where motion is too rapid to allow molecule formation.

Survey

When a theoretical model is proposed for some physical system (as our F_3 force law, for a diatomic molecule), the main questions are: (1) what kinds of motion are allowed mathematically by the theoretical model? and (2) do they correspond to some or all of the motions in the physical system? For the molecular model, the second question is too difficult at this stage of our physics study (one "sees" internal motions in molecules via spectroscopy and quantum mechanics, which we are working towards), though it could be approached for the magnetic-force model, F_4. So we focus here on the first question.

The Phase Plane   To describe the motion (under the F_3 force law) for every possible (x_{init}, v_{init}) is a major challenge. Plotting the motion in the xv plane (or **phase plane**, as it is called) is an important tool.

Each point on a solution curve in the xv plane is a possible initial state for the system. Associated with it is all the information, except the time t, in one row of our spreadsheet calculation. Therefore, the next point on the curve, dt later, is computable from the present point, and this computation is just what the next row in the worksheet does. To show this emphatically, erase the time column (/reK44..K144) from your CBOND worksheet and recalculate. The V_X phase-plane plot is still there! You can also change initial conditions and use {Calc} and {Graph} to explore new cases. (See Exercise 5.4.) There is even time information in the V_X graph if you format the curve as suggested to show a tic mark for each dt step. Thus a single solution curve in the xv plane gives information about choosing any point on it as an initial-condition set (x_{init}, v_{init}). Another helpful feature of the xv plane is that two solution curves can never have a point in common without agreeing everywhere (since each point determines the next, just as each worksheet row determines the next). So solution curves can never cross. These conclusions all follow from the force's being specified as $F(x)$ without an explicit time dependence and would also hold for a velocity-dependent force $F(x, v)$ so long as (x, v) determined the force without any need to know t.

Wholesale Classification   Because solution curves in the xv plane cannot cross, we can categorize entire areas by finding a single closed curve. For example, choosing $v_{init} = 0$ and $x_{init} = x_0 > 1$ should produce a loop enclosing the equilibrium point, $(x, v) = (1, 0)$. (Try it, Exercise 5.5.) Once the curve has closed on itself by returning to its starting point, we know it can only repeat itself if we follow farther forward or backward in time, so it is a periodic bound orbit. Moreover, the points inside this loop must also lie on bound orbits since the solution curves we could draw by starting inside cannot cross the loop to reach larger values of $|x|$ and are therefore bounded. A similar but more complicated argument uses special features of the F_3 function to show that any point outside any scattering orbit is also on a scattering orbit (Exercise 5.6).

Time-Reversal Symmetry   Another helpful property of laws of force of the form $F(x)$ is the **time-reversal invariance** of their solutions. A simple formal calculus calculation shows that if $x = f(t)$ solves $d^2x/dt^2 = F(x)/m$, then so do $x = f(-t)$ and $x = f(t_0 - t)$. This implies symmetries (in the graphs we have created) if the solutions are continued far enough. For example, starting at some $(x, v) = (x_0, 0)$, look at one or more of the graphs we have designed. Then rerun the computations after changing the sign of dt. You should find a mirror

image of the first calculation. In the **V_X** graph, the mirroring is reflection in
the x-axis; in the **X_T** graph, it is reflection in the vertical axis. In general,
the time-reversed solution begins at $(x_0, -v_0)$ if the original solution began at
(x_0, v_0). As a consequence of this symmetry, you need follow a solution only
in the upper half of the xv plane to know how the lower half-plane looks.

Dimensional Analysis   What time unit are we using in these calculations?
Mass and length units were easier—they were the mass of the (single) particle
we follow and a distance that makes the force vanish (in the F_3 and F_4 examples). But a natural time scale is not manifest in the force laws used here to
define examples of motion. Instead it appears when one watches the resulting
motion, e.g., by solving Newton's laws and graphing the results. Now that this
has been done, we can identify in each case a characteristic time interval. The
natural unit of time is influenced both by the mass and the strength of the
force; a smaller mass or a stronger force would lead to higher accelerations,
i.e., to changes occurring on a shorter time scale.

For forces that have an equilibrium (no motion) position where $F = 0$,
a convenient time standard is the period of small oscillations (vibrations,
bounded repetitive motions) about that equilibrium position. In the **CBOND**
worksheet you can find such small oscillations by choosing initial conditions
with $x_{\text{init}} = 1$ and $|v_{\text{init}}| \ll 1$. In Exercise 5.3 you are asked to verify by numer-
ical experiments that these small oscillations have a period of 2π. The **period**
of a motion is the shortest time interval after which (x, v) return to their initial
values. Thereafter the prior motion repeats. The time unit used in the **CBOND**
computations is therefore $(2\pi)^{-1}$ times the period of small oscillations. You
will find this same choice of time unit is implied also by the F_2 and F_4 force
laws suggested in this chapter.

As an example, suppose a molecular spectroscopist asserts that small vibrations in N_2 and O_2 molecules would have periods of, respectively, 1.4 and
2.1 times 10^{-14} s. To treat this worksheet as a model of motion in one of these
molecules, then, you would interpret the graphs as drawn using the corresponding time scale. In fact, the spectroscopist got these numbers for the periods by
comparing a somewhat more elaborate model to the vibrations seen in a real
gas and selecting a time scale that best fits the data. (The data available for
this fit were roughly the energies for oscillations of various discrete sizes, where
size is measured by the area of the loop in the phase plane that the oscillatory
motion describes.)

EXERCISE 5.1^{C} Construct the worksheet **CBOND** shown in Screen 5.1 by modifying
the **LEAP** worksheet. Use these large-scale editing commands before retyping the
changed text and revising the acceleration formula.

> **A9:** /Worksheet Insert Row {Enter}
> **A3:** /Worksheet Delete Row {Enter}
> **E9:** /Range Erase E9..F11
>> Clear unneeded reports and make a new row (8 in Screen 5.1) to display
>> the force law.

C13: U @PI/10   READY

	A	B	C	D	E	F	G	H
1	your name	&	today's date					
2	your class & section			CBOND, model of chemical force				
3					Leapfrog integration			
4	--							
5	Construction parameters:							
6		None. Use atom's mass as mass unit; length and time						
7		units are implied by the force law.						
8			F_3 = (1/x^5 - 1/x^3)/2					
9	Initial data:							
10		x_init =	1 L					
11		v_init =	0.05 L/T					
12	Approximation:							
13		dt =	0.314159 T					
14	--							
15	Notes page:							

M45: +VHS+N45*$DT   EDIT
+M44+N45*$C$13

	I	J	K	L	M	N	O	P
41	name:		time	pos'n	vel_n+1/2	accel	vel_n	
42	units:		[T]	[L]	[L/T]	[L/T^2]	[L/T]	
43	labels:		t	x	vhs	a	v	
44	init data:		0	1	0.05	0	0.05	
45	typical row:		0.314159	1.015707	0.045399	-0.01464	0.047699	
46	copied rows:		0.628318	1.029970	0.037154	-0.02624	0.041276	

Screen 5.1: Integration of the F_3 force motion in CBOND.

As deliverable output of this exercise, produce a {PrintScreen} copy (like Screen 5.1) of your screen layout and a listing of your principal formulae (corresponding to rows 43..46).

EXERCISE 5.2[C]   Find a bound state and a scattering state. Print a V_X graph of each and mark on the printout the initial condition (circle the point in pen) and the category (bound or scattering).

EXERCISE 5.3[C]   Find two sets of initial conditions (x_{init}, v_{init}) with $x_{init} = 1$ where the v_{init} differ by at least a factor of ten and where each is small enough that the periods of oscillation are the same in both cases. Use a time step dt that is a simple fraction of π, such as $\pi/25$, so that the 100 time steps we plot in CBOND add up to an integer multiple of π. Mark (in pen or pencil) on a V_T graph of each case an interval of time that is one period of the motion and give its length as a multiple of π.

EXERCISE 5.4[P]   Use the CBOND worksheet (with its time column intact) to see that the time values (as distinct from time differences) are insignificant. Look at the V_T graph for initial conditions $(x_{init}, v_{init}) = (1, 0.3)$ and print several different cases with only t_0 in cell K44 changed (e.g., $t_0 = 0$, 15, and -785) to show that nothing on the graph changes except the labels on the t-axis.

EXERCISE 5.5^{P} Print a V_X graph of a motion (with force F_3) that includes a
point $(x, v) = (x_0, 0)$ with $x_0 > 1$ and forms a closed loop. Can you find any curve
containing such a point that you think might not close if numerical difficulties (dt
too small or large) didn't prevent continuing it?

EXERCISE 5.6^{E} Find a scattering orbit with the F_3 force law by choosing initial
conditions $(x_{\mathrm{init}}, v_{\mathrm{init}}) = (x_0, 0)$ with a suitable $x_0 < 1$. After getting the solution
started with dt small enough to give believable accuracy, continue it with larger time
steps by using the /Range Value command to copy the last values of x and v to the
initial-value cells. The result should be a loop enclosing the x-axis portion above x_0
as far as you have patience to explore. Argue, given that $F_3(x) > 0$ for $0 < x < 1$
and given the noncrossing of orbits, that all points outside your loop with $x > 0$ lead
to scattering orbits.

Chapter 6

Work and Energy, Dissipation and Conservation

6.1 Energy as a Tool for Understanding Motion

So far we have used kinematic variables (position, velocity, and acceleration) to describe the motion of simple systems. The disadvantage of such description is its completeness—it gives all the information without digesting it. The concept of energy picks out a particularly meaningful subset of this information. In some cases, as in the problems found in most texts, this concept allows one to calculate by hand some features of motion in situations where a computer would be required to calculate the full details. In other cases, as in a gas of interacting particles, not even a computer can calculate the detailed motions, but energy remains a useful summary property.

An important energy insight is the distinction between forces that conserve mechanical energy and those that dissipate it, a distinction illustrated by examples in this chapter.

In computational physics, an important further use of energy is as an error alarm in calculations with conservative forces. If the forces are known to be conservative (by analytic arguments using calculus), then any failure of energy conservation in the computation must be caused either by errors in the formulation or by inadequate accuracy in the numerical approximation. Calculating the energy to check the degree of its constancy is a useful aid in making a valid computation.

6.2   Energy Review

Recall from your main text the important definitions of work, kinetic energy, and (for conservative forces) potential energy.

The work done by the net external force $\mathbf{F}(\mathbf{r})$ acting on a moving particle is

$$W = \int_{t_0}^{t_n} dW \qquad \text{where} \qquad dW = \mathbf{F}(\mathbf{r}) \cdot d\mathbf{r} \quad . \tag{6.1}$$

The particle's kinetic energy at any instant is related to its mass m and its velocity $\mathbf{v}$ by

$$T = \tfrac{1}{2} m \mathbf{v} \cdot \mathbf{v} \quad . \tag{6.2}$$

These two key concepts are related by the work-energy theorem. It states that the work done on the particle during any interval equals the particle's gain in kinetic energy during the interval, or

$$W = T_n - T_0 \quad . \tag{6.3}$$

This theorem is proven analytically by differentiating equation 6.3 with respect to t_n. Illustrating it numerically, however, leads us to further concerns about accurate approximations which are taken up in Section 6.4.

In general the integrand dW in equation 6.1 is not the differential of a function in spite of the notation and might better be written $\tilde{d}W$ to indicate this distinction. A force $\mathbf{F}$ for which the work integrand $\mathbf{F} \cdot d\mathbf{r}$ is the differential of a function is called **conservative**. In this case the function is written $-U(\mathbf{r})$ and U is called the potential energy. Then the property defining conservative,

$$\mathbf{F} \cdot d\mathbf{r} = -dU \tag{6.4}$$

(where U is a function only of the position and does not depend explicitly on time or velocity), can be read as stating that the potential energy decreases by the amount of work done. Since, by the work-energy theorem, this work increases the kinetic energy by the same amount, the total energy $T + U$ will not change. In brief, total energy is not changed by the action of conservative forces.

EXERCISE 6.1^{C} Find functions $U_1(x)$ and $U_2(x)$ that are potential energies for the forces F_1 and F_2 from page 72. Show (by differentiation) that

$$U_3(x) = \frac{1}{8x^4} - \frac{1}{4x^2} \quad , \tag{6.5}$$

i.e., that $F_3 = -dU_3/dx$ agrees with equation 5.5.

6.3   Testing Numerical Approximations

Since energy as an error alarm is among the simplest and most useful of energy applications, we treat it first. An energy column can be added to the CBOND worksheet (or to the ITEST worksheet) and graphed to display the degree of its constancy. A similar check should be made for any motion controlled by a conservative force. A suitable energy formula to use in these worksheets is

$$E_n = \tfrac{1}{2}m{v_n}^2 + U(x_n) \tag{6.6}$$

for the total energy at time t_n. But our use of finite time steps dt is only an approximation to the derivatives that occur in Newton's laws and in the proof of energy conservation. It should not be surprising, then, that other approximations are possible. When the leapfrog integration method is being used, another approximation to $E = \tfrac{1}{2}mv^2 + U(x)$ is

$$E_n = \tfrac{1}{2}mv_{n-1/2}v_{n+1/2} + U(x_n) \quad . \tag{6.7}$$

EXERCISE 6.2[C]   Add an energy column in the work block of each the ITEST and the LEAP worksheets. Then, for the same initial conditions and time step dt in each of these worksheets, print graphs of energy versus time. In this example, what are the principal differences in the energy behavior between the Euler and leapfrog integration schemes?

EXERCISE 6.3[C]   Add an energy column to the CBOND worksheet and add a named graph plotting energy as a function of time. Explore different orbits (small oscillations, large oscillations, scattering) and for each print both the energy graph and a graph of velocity versus time. Under what conditions is it most difficult to conserve energy in the numerical integration?

EXERCISE 6.4[E]   In the LEAP worksheet, use the approximation 6.7 in an energy column and show by examples that energy conservation is exact to the accuracy limit of the computer.

EXERCISE 6.5[H]   Prove analytically that the energy defined by equation 6.7 is conserved exactly (i.e., show that $E_{n+1} - E_n = 0$ with a finite nonzero dt) when the leapfrog integration scheme of equations 4.3 (page 58) is used with a linear force law: $F(x) = -mg - m\omega^2 x$.

6.4   ♠ RK2: Runge-Kutta Integration

To explore how energy ideas can help us understand motion, we return to Section 3.1 and the worksheet DRAG1. We want to modify it to include work and energy calculations alongside the kinematic description already prepared. As we discovered in the last chapter, however, the simple Euler integration scheme used in this sheet does not produce very accurate values. So we should use a better scheme. Accuracy problems here are reduced by an improved integration scheme that (unlike leapfrog) works also with velocity-dependent forces. Such forces arise not only in friction and air drag but also in the important conservative force of magnetism, which we will meet later.

The improved integration scheme we use is called second-order Runge-Kutta, or RK2 for short. (It is called second order because the errors over

labels:	t	h	v	a
init	t_0:	h_0:	v_0:	a_0:
data:	0	h_{init}	v_{init}	$F_0/\mathbf{m}$
typical	t_1:	h_1:	v_1:	a_1:
line:	$t_0 + \mathbf{dt}$	$h_0 + v_{1/2}\,\mathbf{dt}$	$v_0 + a_{1/2}\,\mathbf{dt}$	$F_1/\mathbf{m}$

	hhs	vhs	ahs
	$h_{1/2}$:	$v_{1/2}$:	$a_{1/2}$:
	$h_0 + v_0\,\mathbf{dt}/2$	$v_0 + a_0\,\mathbf{dt}/2$	$F_{1/2}/\mathbf{m}$
	$h_{3/2}$:	$v_{3/2}$:	$a_{3/2}$:
	$h_1 + v_1\,\mathbf{dt}/2$	$v_1 + a_1\,\mathbf{dt}/2$	$F_{3/2}/\mathbf{m}$

Schema 6.1: Second-order Runge-Kutta (RK2) numerical integration. Here the accelerations must be computed from the force law using $F_n = F(x_n, v_n, t_n)$. In the RK2 worksheet, they have formulae such as $a_{3/2} = F_{3/2}/m = -\mathbf{g} - \mathbf{g}v_{3/2}|v_{3/2}|/\mathbf{v}_{\text{term}}^2$.

a fixed time interval $0 \le t \le T$ are proportional to dt^2 rather than being proportional to dt as they were seen to be with the Euler method in Figure 4.3.) The idea behind RK2 integration is quite simple. You merely use the Euler method to obtain v and x at $t + dt/2$ from a, v, and x at t; next calculate a at $t + dt/2$ from these values of v and x; and then use the values of a and v at $t + dt/2$ to leap from t to $t + dt$. Schema 6.1 summarizes the relations among the sheet's cells for the RK2 integration method. Screen 6.1 shows the work block of DRAG1 modified for RK2 integration.

With these preliminaries out of the way, the next section will test how well our numerical approximation fits our expectations based on the work-energy theorem.

EXERCISE 6.6^{C} You should now modify your version of DRAG1 from Exercise 3.1 for RK2 integration, using Schema 6.1 as your guide and Screen 6.1 to check your results. To position the work block as shown here, use the commands

```
I35: /Worksheet Insert Row I35..I40{Enter}
K41: /Range Name Create #{Enter} K41{Enter}
K42: /Worksheet Column Set-width 6{Enter}
J42: /Worksheet Column Set-width 3{Enter}
```

which assume that you had a blank column to the left of the time column, as we did in Screen 2.1. You may notice that all the numbers in Screen 6.1 are rounded to two decimal places. You can modify your sheet to provide this easy-to-read output by typing /Worksheet Global Format Fixed 2{Enter} to change the global format to fixed, rounded to two decimal places. Many alternative formats are available in the same menu. Also, you will find it handy to use the /Range Name Labels Down command again to associate the name of each variable in the work block of your sheet

P45: +V+A*$DT/2   `READY`

	I	J	K	L	M	N	O	P	Q
41					At time t			At time t+dt/2	
42	name:		time	height	velocity	accel	height	velocity	accel
43	units:		[s]	[m]	[m/s]	[m/s^2]	[m]	[m/s]	[m/s^2]
44	labels:		t	h	v	a	hhs	vhs	ahs
45	init data:		0.00	0.00	60.00	-49.00	3.00	57.55	-45.86
46	typical row:		0.10	5.76	55.41	-43.24	8.53	53.25	-40.68
47	copied rows:		0.20	11.08	51.35	-38.51	13.65	49.42	-36.39

Screen 6.1: The work block of RK2, a worksheet you can use as a template
for any problem needing Runge-Kutta integration. Note that the initial
data needed to produce these results are shown in appropriate cells in
the $t = 0$ row here, where formulae have reproduced them from the I/O
page.

with the numerical value immediately below it. After completing these changes and
rewriting the range-name table at A18, save your sheet under the new name RK2.

6.5 ♠ DRAGW: Testing the Work-Energy Theorem

The DRAG1 worksheet described vertical motion of a projectile in a uniform
gravitational field with an added force proportional to the projectile's velocity
squared, directed opposite to its velocity vector. This v-squared force models
the air resistance acting on the projectile. In the version of DRAG1 in Screen 2.1
(page 33), the cursor has been positioned on cell F18 (named A) at the head
of the acceleration column to display the acceleration formula. (After the
reorganization in Section 3.1 and Exercise 6.6, this has become cell N45 as in
Screen 6.1, but still has the name A.) Take a minute to examine this formula and
convince yourself that it correctly describes the acceleration of our projectile.

To check the work-energy theorem, we insert three new columns into the
work block of a renamed RK2, just to the right of column N headed `accel` (for
time t, not $t + dt/2$).

> --: /File Save DRAGW{Enter}
> O44: /Worksheet Insert Column {Right 2} {Enter}
>> You should see highlighted a range three columns wide, beginning
>> in column O, before you complete this command. Its contents will
>> be pushed to the right as three empty columns are inserted.

The first of these three new columns will tabulate the work W done by the net
force on the projectile. The next will tabulate the projectile's kinetic energy
T. The third will prepare for a graph to check the work-energy theorem. If the

labels:	W		KE	Delta
init	W_0:		T_0:	Δ_0:
data:		0	$mv_0^2/2$	$T_0 - W_0$
typical	W_1:		T_1:	Δ_1:
line:	$W_0 + ma_{1/2}(h_1 - h_0)$		$mv_1^2/2$	$T_1 - W_1$

Schema 6.2: These are the formulae needed to calculate the work done by the net force acting on the projectile and to compare the work done and the resulting change in the projectile's kinetic energy.

theorem holds in our numerical approximation, then, as positive work is done, kinetic energy will correspondingly increase and the difference $\Delta = T - W$ will remain constant. We therefore calculate and graph this difference. Schema 6.2 summarizes the formulae needed to calculate these quantities, and Screen 6.2 displays the column headings and the first few numerical values. In adding these quantities to your sheet, you will need to add a row to the I/O page to specify the mass of the projectile, which we will take to be 1 kg.

> **A9:** `/Worksheet Insert Row {Enter}`
> **D2:** `DRAGW: v-squared air resistance`
> **D3:** `/Worksheet Delete Row {Enter}`
> **C8:** `/Range Name Create M{Enter} {Enter}`
>> Provide a new input data row for mass, as seen in Screen 6.2, and revise the main title.

It is also useful to be sure that all the range names are still correct and to add a summary result computation:

> **K44:** `/Range Name Labels Down K44..T44{Enter}`
> **Q45:** `/Range Name Create DELTA{Enter} Q45..Q145{Enter}`
> **E12:** `Result:`
> **E13:** `"W_err =`
> **F13:** `(@max(delta)-@min(delta))/KE`
> **F13:** `/Range Format Percent 4{Enter} {Enter}`
>> By choosing the entire column of Δ values as a named range, we can write a more understandable formula in the report cell **F13** that gives the percentage variation in Δ, which should ideally be zero.

EXERCISE 6.7[C] Incorporate the preceding changes into the sheet RK2 and save your sheet, with the new name DRAGW, for later modifications. Explore the KE-W column to find the largest and the smallest values of this supposed constant. While you are doing this, it may be convenient to keep the column headings in view:

> **Q46:** `/Worksheet Titles Horizontal`

To leave the work block, cancel this configuration with the command `/Worksheet Titles Clear`.

F13: (P4) (@MAX(DELTA)-@MIN(DELTA))/KE   **READY**

	A	B	C	D	E	F	G	H
1	your name	&	today's date					
2	your class & section			DRAGW: v-squared air resistance				
3					Runge-Kutta integration			
4	--							
5	Construction parameters:							
6		g =	9.80 m/s^2		Gravity.			
7		v_term =	30.00 m/s		Terminal velocity.			
8		mass =	1.00 kg		Projectile mass.			
9	Initial data:							
10		h_init =	0.00 m		Initial altitude.			
11		v_init =	60.00 m/s		Initial velocity.			
12	Approximation:				Results:			
13		dt =	0.10 s		W_err =	0.1775%		

	I	J	K	L	M	N	O	P	Q
41				At time t					
42	name:	time	height	velocity	accel	work	K.E.	KE-W	
43	units:	[s]	[m]	[m/s]	[m/s^2]	[J]	[J]	[J]	
44	labels:	t	h	v	a	W	KE	Delta	
45	init data:	0.00	0.00	60.00	-49.00	0.00	1800.00	1800.00	
46	typical row:	0.10	5.76	55.41	-43.24	-263.95	1535.33	1799.28	
47	copied rows:	0.20	11.08	51.35	-38.51	-480.57	1318.19	1798.76	

Screen 6.2: Two segments from DRAGW, including a new row specifying the mass of the projectile and three new columns used to apply the work-energy theorem to this system.

EXERCISE 6.8[P]   Using DRAGW, explore how the numerical error you have in Δ depends on the size of the time step dt. A simple way to do this is to cut the time step in half and then double the number of rows in the work block so that the total elapsed time remains constant. A second-order integration method, such as RK2, should reduce the error to $1/2^2 = 1/4$ of its former value when the integration step size is cut in half. Is this true for your calculation?

EXERCISE 6.9[C]   From DRAGW, make a graph of the difference $\Delta = T - W$ as a function of time, t. Repeat for Δ versus h and for Δ versus v. Using these three graphs, explain how the numerical error resulting from the finite integration-step size depends on the variables of this problem.

EXERCISE 6.10[E]   The RK2 method of solving Newton's law gives sufficient data to do an even more accurate evaluation of the work integral than we have used in Schema 6.2. Simpson's rule for numerically evaluating integrals assigns relative weights 1,4,1 to the beginning, middle, and end of an interval. (It is based on approximating by a quadratic polynomial the function to be integrated—see almost any calculus text.) In Schema 6.2 it requires replacing the formula for W_1 with $W_1 = W_0 + m(a_0 + 4a_{1/2} + a_1)(h_1 - h_0)/6$. In column O of DRAGW.wk1 modify the formulae that calculate the work done so that they use this formula and then compare the Δ variation in this case with similar results from the previous method.

	K	L	M	N	O	P	Q	R
42	time	height	velocity	accel	work	K.E.	P.E.	total E
43	[s]	[m]	[m/s]	[m/s^2]	[J]	[J]	[J]	[J]
44	t	h	v	a	W	KE	PE	E
45	0.00	0.00	60.00	-49.00	0.00	1800.00	0.00	1800.00
46	0.10	5.76	55.41	-43.24	-263.95	1535.33	56.40	1591.73
47	0.20	11.08	51.35	-38.51	-480.57	1318.19	108.59	1426.78
48	0.30	16.02	47.71	-34.58	-660.43	1137.95	157.02	1294.96
49	0.40	20.62	44.42	-31.29	-811.32	986.77	202.08	1188.84

Screen 6.3: A portion of the DRAGW work block in which the potential energy of the projectile is combined with its kinetic energy to obtain its total mechanical energy.

6.6 ♠ Conservation of Mechanical Energy

The work-energy theorem is true for any combination of external forces doing work on an object. It does not matter whether the forces are conservative, like gravity, or dissipative, like air drag. The difference between interactions that store energy in a recoverable form and those that convert mechanical energy irreversibly into heat is, however, crucial to understanding how our projectile moves. To see this, we will now insert two additional columns into DRAGW, one to tabulate the gravitational potential energy $U = mgh$ of the projectile and the other listing total mechanical energy $E = T + U$. We insert these two columns immediately to the right of the current kinetic-energy (KE) column. Screen 6.3 shows a portion of the work block that results from this.

EXERCISE 6.11[C]   Insert the two new columns into DRAGW and make a graph E_T of the total energy E as a function of the elapsed time t. The observed loss of energy is, of course, caused by the dissipative air drag experienced by the projectile during its flight, which a graph E_V of total energy E versus $|v|$ can help illustrate. Save your worksheet including these two new graphs.

EXERCISE 6.12[C]   The strength of the air drag is determined by the value of the terminal speed, v_{term}, set on the I/O page of DRAGW. Explore the effect of air drag on the motion and energy of the projectile by setting v_{term} first to smaller values and then to much larger values. For each extreme, document with graphs how the important variables behave in the limit of large and small drag. In particular, find the value of v_{term} that limits the loss of total energy to 1% of its initial value during the first 10 seconds of the projectile's motion.

EXERCISE 6.13[P]   What effect does reducing the air-drag force have on the numerical error present in Δ?

6.7   Listings

```
range names:
#       I41     T       K45
A       N45     TITLE1  D2          I42: 'name:         K42: [W6] ^time
AHS     Q45     TITLE2  A1          I43: 'units:        K43: [W6] ^[s]
DT      C13     V       M45         I44: 'labels:       K44: [W6] ^t
G       C7      VHS     P45         I45: 'init data:    K45: [W6] 0
H       L45     V_INIT  C11         I46: 'typical row:  K46: [W6] +T+$DT
HHS     O45     V_TERM  C8          I47: 'copied rows:  K47: [W6] +K46+$DT
H_INIT  C10

L41: '      At time t
L42: ^height        M42: ^velocity    N42: ^accel
L43: ^[m]           M43: ^[m/s]       N43: ^[m/s^2]
L44: ^h             M44: ^v           N44: ^a
L45: +H_INIT        M45: +V_INIT      N45: -$G-$G*(V/$V_TERM)*@ABS(V/$V_TERM)
L46: +H+VHS*$DT     M46: +V+AHS*$DT   N46: -$G-$G*(M46/$V_TERM)*@ABS(M46/$V_TERM)
L47: +L46+P46*$DT   M47: +M46+Q46*$DT N47: -$G-$G*(M47/$V_TERM)*@ABS(M47/$V_TERM)
...
L145..N145

O41: '      At time t+dt/2
O42: ^height        P42: ^velocity    Q42: ^accel
O43: ^[m]           P43: ^[m/s]       Q43: ^[m/s^2]
O44: ^hhs           P44: ^vhs         Q44: ^ahs
O45: +H+V*$DT/2     P45: +V+A*$DT/2   Q45: -$G-$G*(VHS/$V_TERM)*@ABS(VHS/$V_TERM)
O46: +L46+M46*$DT/2 P46: +M46+N46*$DT/2 Q46: -$G-$G*(P46/$V_TERM)*@ABS(P46/$V_TERM)
O47: +L47+M47*$DT/2 P47: +M47+N47*$DT/2 Q47: -$G-$G*(P47/$V_TERM)*@ABS(P47/$V_TERM)
...
O144..Q145
```

Listing 6.1: Listing of the RK2 worksheet.

```
range names:
#       K41     H       L45     TITLE1  D2
A       N45     HHS     T45     TITLE2  A1
AHS     V45     H_INIT  C10     V       M45       E12: U 'Results:
DELTA   S45..S145  KE   P45     VHS     U45       E13: "W_err =
DT      C13     M       C8      V_INIT  C11       F13: (P4) (@MAX(DELTA)-@MIN(DELTA))/KE
E       R45     PE      Q45     V_TERM  C7
G       C6      T       K45     W       O45

O42: ^work              P42: ^K.E.      Q42: ^P.E.      R42: ^total E  S42: ^KE-W
O44: ^W                 P44: ^KE        Q44: ^PE        R44: ^E        S44: ^Delta
O45: 0                  P45: +$M*V^2/2  Q45: +$M*$G*H   R45: +KE+PE    S45: +KE-W
O46: +W+$M*(L46-H)*AHS  P46: +$M*M46^2/2 Q46: +$M*$G*L46 R46: +P46+Q46 S46: +P46-O46
O47: +O46+$M*(L47-L46)*V46 P47: +$M*M47^2/2 Q47: +$M*$G*L47 R47: +P47+Q47 S47: +P47-O47
...
O145..S145
```

Listing 6.2: Listing for the DRAGW worksheet.

Part II

Applications

Chapter 7

Torque and Angular Momentum in the Simple Pendulum

The principal worksheet studied in this chapter is PEND, described in Section 7.2. It follows the motion of a simple pendulum without making any small-angle approximation. A simple pendulum is a rigid massless rod with a point mass at one end and the other end attached to a frictionless pivot allowing the rod to rotate freely in a vertical plane. If you have already studied torques and angular momentum, you may skip directly to Section 7.2 and begin on the worksheet. For students in courses that have omitted the study of rigid bodies in rotation, Section 7.1 introduces torque and angular momentum in the context of particle motion. It is needed to find the equation of motion of a simple pendulum. The final section (7.3) reiterates the physicist's idea that to "solve a problem" means to create a conceptual model (here of a pendulum) and to explore the model's behaviors and their relationships to physical phenomena.

7.1   Torque and Angular Momentum

The simple pendulum looks like an elementary mechanical system. We will solve it using Newton's theory that $\mathbf{a} = \mathbf{F}/m$, but we can't use the theory directly. A difficulty appears when we draw a free body diagram for the mass at the moving end of the rod. The forces on it are $m\mathbf{g}$ (downward) and a contact force $\mathbf{T}$ (from the rod). $\mathbf{T}$ is directed inward (toward the pivot) when the tension T in the rod is positive and outward when the rod is being compressed ($T < 0$). The difficulty in using Newton's laws to find the acceleration $\mathbf{a} = (a_x, a_y)$ of the mass's position $\mathbf{r} = (x, y)$ is that the tension is not known *a priori* (beforehand). Exercise 7.3 suggests you use the fact that $r^2 = x^2 + y^2$

is constant (because the rod is stiff) to obtain by differentiation an equation that yields T. In this section, we will attack the T difficulty with a different strategy, which introduces concepts (torque and angular momentum) that have many other applications. The strategy is to find a combination of the a_x and a_y equations giving the acceleration of the angular position $\theta(t)$ that interests us—while ignoring the radial position $r(t) = R = \text{const}$, which serves only to determine T.

Let us begin by looking at velocity $\mathbf{v}$ and momentum $\mathbf{p} = m\mathbf{v}$ to isolate their radial and transverse parts. The dot product $\mathbf{r} \cdot \mathbf{v} = rv_r$ picks out the radial component of velocity v_r. In the case of the simple pendulum, it is zero from

$$\mathbf{r} \cdot \mathbf{v} = x\frac{dx}{dt} + y\frac{dy}{dt} = \frac{1}{2}\frac{d}{dt}(x^2 + y^2) = \frac{d}{dt}(\frac{1}{2}r^2) \tag{7.1}$$

since $r^2 = \text{const}$. The vector cross product $\mathbf{r} \times \mathbf{v}$ or $\mathbf{r} \times \mathbf{p}$ will ignore any radial component of velocity or momentum. To see this, suppose that the momentum $\mathbf{p} = m\mathbf{v}$ contains a radial part $(\mathbf{r}/r)p_r$ in the $\mathbf{r}$ direction so that $\mathbf{p} = \mathbf{p}_\perp + \mathbf{r}(p_r/r)$, where $\mathbf{p}_\perp$ is the remaining transverse part. Then calculate

$$\mathbf{r} \times \mathbf{p} = \mathbf{r} \times \mathbf{p}_\perp + \mathbf{r} \times \mathbf{r}(p_r/r) = \mathbf{r} \times \mathbf{p}_\perp \tag{7.2}$$

since $\mathbf{r} \times \mathbf{r} = 0$. Thus any radial part of $\mathbf{p}$ disappears from the product $\mathbf{r} \times \mathbf{p}$, and we have in $\mathbf{r} \times \mathbf{p}$ a quantity sensitive only to the transverse part of $\mathbf{p}$. (Cross products may be new to you. Review the definitions in your text and check that $\mathbf{A} \times \mathbf{A} = 0$ for any vector $\mathbf{A}$. This also follows from the identity $\mathbf{A} \times \mathbf{B} = -\mathbf{B} \times \mathbf{A}$ applied to the case $\mathbf{B} = \mathbf{A}$.)

Radial forces such as $\mathbf{T}$ in our pendulum problem (or the sun's gravitation in the solar system) cause changes in the radial momentum only and get special attention because there is a preferred center, $r = 0$, in many problems. The transverse momentum, which such forces cannot change, is analyzed by isolating it in $\mathbf{r} \times \mathbf{p}$, which gets a special name and symbol:

$$\mathbf{L} \equiv \mathbf{r} \times \mathbf{p} \equiv \text{angular momentum.} \tag{7.3}$$

Let us now calculate from Newton's laws how $\mathbf{L}$ does change when forces act on a point mass:

$$\frac{d\mathbf{L}}{dt} = \frac{d}{dt}(\mathbf{r} \times \mathbf{p}) = \frac{d\mathbf{r}}{dt} \times \mathbf{p} + \mathbf{r} \times \frac{d\mathbf{p}}{dt} = \mathbf{r} \times \mathbf{F} \quad . \tag{7.4}$$

(If you are surprised that the product rule for differentiation is still so simple when applied to vector products, you can verify it when calculating a single vector component in equation 7.9.) The last step here uses $d\mathbf{p}/dt = \mathbf{F}$ in the second term and $\mathbf{v} \equiv d\mathbf{r}/dt$ in the first term

$$\frac{d\mathbf{r}}{dt} \times \mathbf{p} = \mathbf{v} \times (m\mathbf{v}) = m(\mathbf{v} \times \mathbf{v}) = 0 \tag{7.5}$$

again using $\mathbf{A} \times \mathbf{A} = 0$ for any vector.

Note that if there is a radial part of $\mathbf{F}$ (in the direction of $\mathbf{r}$), it will not contribute to $\mathbf{r} \times \mathbf{F}$ and therefore not to the change in $\mathbf{L}$. The quantity $\mathbf{r} \times \mathbf{F}$ thus is important and is given a name:

$$\mathbf{N} \equiv \mathbf{r} \times \mathbf{F} \equiv \text{torque.} \tag{7.6}$$

(The Greek tau τ is also used for torque, but $\mathbf{N}$ will be easier to type in our worksheets.) We have thus derived the single-particle case of an important corollary to Newton's law: Torque causes changes in angular momentum, or

$$\frac{d\mathbf{L}}{dt} = \mathbf{N} \quad . \tag{7.7}$$

For the pendulum problem, where all motion is in the xy plane, only the z-component of $\mathbf{L}$ is nonzero. It is given by

$$L_z = (\mathbf{r} \times \mathbf{p})_z = xmv_y - ymv_x \tag{7.8}$$

and the calculation in equation 7.4 becomes

$$\begin{aligned} \frac{dL_z}{dt} &= \frac{dx}{dt}mv_y - \frac{dy}{dt}mv_x + xm\frac{dv_y}{dt} - ym\frac{dv_x}{dt} \\ &= v_x mv_y - v_y mv_x + xF_y - yF_x \\ &= xF_y - yF_x \equiv N_z \quad . \end{aligned} \tag{7.9}$$

Now substitute the forces on the pendulum mass (see Figure 7.1 and note that similar right triangles are formed by xyr and $T_x T_y T$)

$$\begin{aligned} F_x &= -(x/r)T \\ F_y &= -mg - (y/r)T \end{aligned} \tag{7.10}$$

to find

$$N_z = -mgx \tag{7.11}$$

from which T has disappeared. The next section explains how this and the definition of L_z are used to calculate the theoretical motions of a simple pendulum.

EXERCISE 7.1$^\text{C}$ Assume that the radial component F_r of a vector $\mathbf{F}$ is defined by $rF_r = \mathbf{r} \cdot \mathbf{F}$ for any vector $\mathbf{F}$ and that the transverse part $\mathbf{F}_\perp$ is defined as the rest of $\mathbf{F}$ according to $\mathbf{F} = \mathbf{F}_\perp + \mathbf{r}(F_r/r)$. Using $r^2 = \mathbf{r} \cdot \mathbf{r}$, show that $\mathbf{r}/r$ is a unit vector. Show further that $\mathbf{r} \cdot \mathbf{F}_\perp = 0$, i.e., that $\mathbf{F}_\perp$ is perpendicular to $\mathbf{r}$.

EXERCISE 7.2$^\text{P}$ Using the definitions in the previous exercise, calculate the radial component T_r of the force $\mathbf{T} = (-xT/r, -yT/r)$.

EXERCISE 7.3$^\text{E}$ Differentiate $x^2 + y^2 = R^2 = \text{const}$ twice to obtain an equation involving the accelerations a_x and a_y, which you should then evaluate from the forces arising from T and mg (as you did in the evaluation of torque). Solve this equation for the tension T. Derive the same equation directly from $\mathbf{F} = m\mathbf{a}$, using a free body diagram and the idea that (in circular motion) centripetal acceleration is v^2/r.

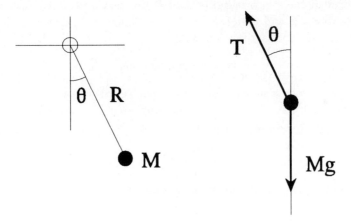

Figure 7.1: Simple pendulum. The sketch on the left shows the geometrical relationship of the frictionless pivot to the moving point mass at the end of the rigid massless rod. On the right is a force diagram showing all the forces acting on the point mass.

7.2   PEND: Simple Pendulum

We want to calculate the motion of a simple pendulum as sketched in Figure 7.1. The pendulum consists of a point mass M at the free end of a rigid massless rod of length R pivoted without friction in a vertical plane. The approach to this problem looks familiar in its outline: we first describe the mechanism as simply as possible (kinematics) and then apply Newton's laws to predict successive increments in momentum and position (dynamics). The new feature of the method is that we use angular positions and momenta. The results allow us to compare the mathematical model with a physical apparatus that is easily constructed and has a variety of motions.

The position of the pendulum will be described by an angle θ measured counterclockwise from its equilibrium position:

$$
\begin{aligned}
x &= R\sin\theta \\
y &= -R\cos\theta \quad .
\end{aligned}
\tag{7.12}
$$

In order that the force exerted radially by the rod on the mass does not enter the calculation, we use the torque equation (7.7) to determine the motion. The force $\mathbf{T}$ exerted by the rod points (for positive T) in the direction opposite $\mathbf{r}$, so $\mathbf{T} = -(\mathbf{r}/r)T$ and gives zero torque: $\mathbf{r} \times \mathbf{T} = -\mathbf{r} \times \mathbf{r}(T/r) = 0$. The total torque, which we write as $\mathbf{N}$, is thus produced entirely by the gravitational force, $M\mathbf{g}$, and has only a z component (perpendicular to the plane of the motion). It can be computed from $N_z = xF_y - yF_x$, with $F_y = -Mg$, $F_x = 0$,

Time	Angle	AngVel	Ang_Mom	Torque
t_0:	θ_0:	$\Omega_{1/2}$:	$L_{1/2}$:	N_0:
0	θ_{init}	$L_{1/2}/(\mathbf{MR}^2)$	$L_0 + N_0\,\mathbf{dt}/2$	$-\mathbf{MgR}\sin\theta_0$
t_1:	θ_1:	$\Omega_{3/2}$:	$L_{3/2}$:	N_1:
$t_0 + \mathbf{dt}$	$\theta_0 + \Omega_{1/2}\,\mathbf{dt}$	$L_{3/2}/(\mathbf{MR}^2)$	$L_{1/2} + N_1\,\mathbf{dt}$	$-\mathbf{MgR}\sin\theta_1$

Schema 7.1: The simple-pendulum worksheet PEND is based on this leapfrog schema. It implements a finite-difference version of equations 7.15. The angles at the integer steps provide torque values defining the changes of angular momentum from one half step to the next. The angular velocities at half-step points define the changes in the angles between integer points.

and is

$$N_z = -x\,Mg = -MgR\sin\theta \quad . \tag{7.13}$$

This result can also be obtained directly from a sketch of the pendulum and the forces on the mass. There we see that the line of action of the gravitational force passes a distance $r_\perp = R\sin\theta$ from the pivot that is our chosen origin and that the rotational sense of this torque is clockwise (negative). The torque gives an equation for dL_z/dt, which must be supplemented by an equation that relates angular momentum L_z to angular velocity $d\theta/dt$ to complete our predictions of the pendulum motions.

By definition, the angular momentum is $\mathbf{L} \equiv \mathbf{r} \times M\mathbf{v}$ and here has only a z component L_z which can be calculated from $L_z = x\,Mv_y - y\,Mv_x$ and equations 7.12 and their derivatives. But it is also easily calculated as $RMv_\perp$, with $v_\perp = R\,d\theta/dt$, giving $L_z = MR^2 d\theta/dt$.

Our two computations expressing torque and angular momentum in terms of θ can now be summarized in two equations

$$\begin{aligned} dL/dt &= -MgR\sin\theta \\ L &= MR^2 d\theta/dt \end{aligned} \tag{7.14}$$

in which, and henceforth, we write L for L_z. The two equations 7.14 are respectively the dynamic-torque equation 7.7 and the definition of angular momentum. As we did in the analogous case for the force equation and the definition of velocity, we use these equations in a spreadsheet to update angular position and momentum according to

$$\begin{aligned} d\theta &= (L/MR^2)dt \\ dL &= -MgR\sin\theta\,dt \quad . \end{aligned} \tag{7.15}$$

This is shown in more detail in Schema 7.1.

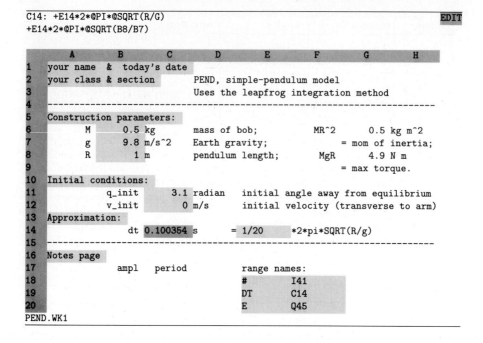

```
C14: +E14*2*@PI*@SQRT(R/G)                                              EDIT
+E14*2*@PI*@SQRT(B8/B7)
```

	A	B	C	D	E	F	G	H
1	your name & today's date							
2	your class & section			PEND, simple-pendulum model				
3				Uses the leapfrog integration method				
4	--							
5	Construction parameters:							
6		M	0.5 kg		mass of bob;		MR^2	0.5 kg m^2
7		g	9.8 m/s^2		Earth gravity;			= mom of inertia;
8		R	1 m		pendulum length;		MgR	4.9 N m
9								= max torque.
10	Initial conditions:							
11		q_init	3.1 radian		initial angle away from equilibrium			
12		v_init	0 m/s		initial velocity (transverse to arm)			
13	Approximation:							
14		dt	0.100354 s		= 1/20	*2*pi*SQRT(R/g)		
15	--							
16	Notes page							
17		ampl	period		range names:			
18					#	I41		
19					DT	C14		
20					E	Q45		

PEND.WK1
```

Screen 7.1: The home screen of the PEND worksheet, which calculates the motion of a simple pendulum. The beginning of the notes page is shown below the I/O page.

Since many aspects of worksheet construction should be familiar to you by now, the worksheet PEND, with the familiar steps already done, is provided in the student software package. Its home screen is shown here as Screen 7.1, which displays both the I/O page and the beginning of a notes page with space to record your exploration of the period of the pendulum as calculated for a variety of amplitudes (maximum $\theta$ values during an oscillation). In the I/O area, you need only supply identifications in cells A1 and A2. In the work block, shown completed in Screen 7.2, the labels are provided, but you are expected to fill in the physics, following Schema 7.1. (In case the student software package is unavailable, more details are given in the listings in Section 7.4.) The work block includes an energy column which will sometimes—when energy fails to be conserved—warn you of bugs in your formulae or of inaccuracies in the numerical approximations required when working with finite (not infinitesimal) steps $dt$. The formula used there is

$$\begin{aligned} E &= Mv^2/2 + Mg(y+R) \\ &= L^2/(2MR^2) + MgR(1-\cos\theta) \end{aligned} \tag{7.16}$$

and, for all the time steps beyond the first, the approximation

$$(L^2)_n = L_{n+\frac{1}{2}} L_{n-\frac{1}{2}} \tag{7.17}$$
```

P45: +M*V_INIT*R   EDIT
+B6*C12*B8

	I	J	K	L	M	N	O	P	Q
41									
42	name:		time	angle	ang vel	ang mom	torque	ang mom	energy
43	units:		[s]	[1]	[/s]	[J s]	[N m]	[J s]	[J]
44	labels:		t	q	whs	Lhs	N	L	E
45	init data:		0	3.1	-0.0204	-0.01022	-0.20374	0	9.795762
46	typical row:		0.1003	3.0979	-0.0633	-0.03167	-0.21379	-0.02095	9.795657
47	copied rows:		0.2007	3.0915	-0.1125	-0.05625	-0.24491	-0.04396	9.795657
48			0.3010	3.0802	-0.1727	-0.08637	-0.30015	-0.07131	9.795657
49			0.4014	3.0629	-0.2500	-0.12500	-0.38489	-0.10569	9.795657
50			0.5017	3.0378	-0.3518	-0.17591	-0.50731	-0.15045	9.795656

Screen 7.2: The beginnings of the work block of the PEND worksheet. The calculations correspond to the input data shown in Screen 7.1.

is used because it leads to a highly constant energy in simple cases. Note that if you add frictional forces to this pendulum model, the energy will no longer be conserved. And if you change the force law, e.g., by approximating $\sin\theta \simeq \theta$, then the energy formulae also should be changed (or energy will remain conserved only when the new force is nearly the same as the old).

EXERCISE 7.4[C]   Complete the PEND worksheet by adding formulae corresponding to Schema 7.1. Use {PrintScreen} to produce a screen image whose top portion corresponds to Screen 7.2 but that includes the additional rows to make a full screen. Save your completed worksheet for further exploration.

EXERCISE 7.5[P]   Modify the PEND worksheet by changing the torque formula to an approximate form useful analytically at small angles, namely $N = -MgR\theta$ (from $\sin\theta \simeq \theta$). For consistency, you should use also an energy formula that matches this torque. From $dU = -F\,ds = -FR\,d\theta = -N\,d\theta$, find the appropriate expression for the potential energy, U. Print graphs showing that the period of oscillation is the same for oscillations with maximum θ values of 0.1, 1, and 10 radians, and explain in a few sentences how wildly the behavior of the pendulum predicted by this (approximate) model differs from that of an actual pendulum in the case computed with $\theta_{max} = 10$.

7.3   Exploring Mathematical Models

Creating a worksheet such as PEND is only the first step in solving a physics problem. It puts our ideas, our conceptual model of the object we are trying to understand, into a concrete mathematical form where the consequences of those ideas can be examined. The ideas we have put into the worksheet are Newton's laws in the $d\mathbf{L}/dt = \mathbf{N}$ form, together with an idealization of a pendulum with a frictionless pivot, a massless rigid rod, and a point mass. The next step, which we take up here, is drawing out the consequences of this mathematical model and comparing them with actual physical pendula.

Pure Exploration

One should not take a worksheet or other substantial calculation seriously without testing it first, but a healthy curiosity will want to try it out quickly. There is always time for more careful testing after you get acquainted, unless some implausible results suggest there may be bugs to squash. A first try here is simply to see if your mathematical pendulum swings back and forth like an actual pendulum. For that, think of an example (such as a grandfather clock) and put in appropriate parameters. You might check that it does swing, i.e., oscillate between positive and negative θ repeatedly, by starting it with an initial angle (called `q_init` in the worksheet) of $\theta_0 = 0.1$ radian, which is about $6°$, and with $v_0 = 0$. For a more thorough exploration, try a variety of initial angles between 0 and 10 radians. Look for qualitative differences in the motion for various ranges of θ_0 and see if you can understand what they mean for a pendulum that may need to be more ideal than the actual one in a grandfather clock. You will, of course, be looking at graphs to inspect the action of your mathematical model; several named graphs are already defined in the distributed worksheet **PEND**.

Another line of exploration is to set θ_0 to zero and vary the initial velocity. For the construction parameters shown in Screen 7.1, the range $0 \leq v_0 < 10\,\mathrm{m/s}$ leads to motions in an energy range not reached by the θ_0 survey. Again, you should be looking for qualitatively different curves on the graphs, trying to associate them with motions of an appropriately well-built pendulum.

Testing

Your worksheet can be tested, in a way not normally available, by comparing it with Screens 7.1 and 7.2. But how did *we* test it? And how should you retest it if you modify it to do something different (say by adding friction or a driving motor)? The common procedure is to use the worksheet in special cases where you can calculate the answers by other means. Do not overlook the simplest cases. In the simple case where the pendulum is hanging straight down and not moving, it should stay that way. Mathematicians will note that these words assert that $\theta(t) = 0 = L(t)$ is a solution of the differential equations 7.14. So if you have not already done so, set these initial conditions and check the result. A lot of bugs, such as forgetting to mark a constant for absolute addressing before copying a formula, can show up in a test as simple as this.

Another test is to reduce this motion to one previously studied and see if the expected results are produced. One simple motion is free fall; to get it here, we must find a case where the pendulum rod neither pushes nor pulls on the mass so that only gravity acts. Try to think of such a situation. It need last only a short time, and it is clearly impossible for any large part of an oscillation since no oscillations occur in free fall. The trick is to start the pendulum with zero velocity (so the rod need supply no centripetal acceleration) at $\theta = \pi/2$. For a short time thereafter, the mass will fall essentially straight down, and any small force the rod exerts will be mostly horizontal, so the vertical motion should act like free fall. If you choose $M = 1$ and $R = 1$, then $\theta - \pi/2$ and L will be numerically equal to y and v_y while $\theta - \pi/2$ is small.

Order of Magnitude

How small should dt be to keep numerical approximation errors reasonable? More generally, in this problem what is a small energy? or angle? or initial velocity? This kind of question in physics always leads first to dimensional analysis.

For angles the answer is easy. Since angles are dimensionless, a radian is a moderate sized angle and $\theta \ll 1$ is small. (From the geometry of a circle, we know that $|\theta| \leq \pi$ is sufficient for any position description, although sometimes we record some position history as well by using larger angles.)

For time scales we look among the quantities that define our pendulum (M, R, and g) and its initial state of motion (θ_0 and v_0) and search for quantities dimensioned as time. Only g and v_0 involve the unit of time. By combining them with R to eliminate the length unit, we find two time scales that can be written as frequencies (reciprocal times):

$$
\begin{aligned}
\omega_0 &= \sqrt{g/R} \\
\Omega_0 &= v_0/R = (d\theta/dt)_0 \quad .
\end{aligned}
\tag{7.18}
$$

One expectation, then, is that both $\omega_0\, dt$ and $\Omega_0\, dt$ must be much less than unity for dt to be considered small. Any computation that violates one of these conditions should be considered suspect. The I/O screen of PEND simplifies the choice of dt by making dt proportional to $1/\omega_0$. Another implication is that the behavior of the pendulum can change depending on the size of the initial angular velocity of the mass (in the sense that Ω_0/ω_0 is big or small compared to 1). Said another way, a change in behavior (with, say, $\theta_0 = 0$) might be expected for some critical initial velocity, $v_0 = R\Omega_0$, around the value $R\omega_0 = \sqrt{gR}$. Dimensional analysis cannot say whether any such effect occurs at half this value, or π times it, or some other value.

Energy considerations offer additional insight into the behavior of the physical pendulum. There is a natural unit of energy in this problem, $MgR = MR^2\omega_0^2$. You should explore initial conditions that give initial energies that are both low (0.2) and high (5) multiples of this unit and then find the dividing line in energy that marks the change from one type of behavior to the other.

EXERCISE 7.6[C]   Choose initial conditions $v_0 = 0$ and $\theta_0 = 0.1$ radian, then print out a graph of $\theta(t)$ showing normal pendulum oscillations. Explain the differences between this plot and the one that results from the initial conditions in Screen 7.1.

EXERCISE 7.7[P]   Survey a number of initial conditions with $v_0 = 0$ for θ_0 in the range 0.01 to 10 radians. Find critical values of θ_0 that show extreme qualitative differences in the $\theta(t)$ plots. Print graphs of these critical cases and interpret them through annotations in nonmathematical language. Include notes giving the energy of each case in units of MgR.

EXERCISE 7.8[C]   Survey a number of initial conditions with $\theta_0 = 0$ for v_0/R in the range 0 to $4\omega_0$. Find critical values of energy (in units of MgR) that show extreme qualitative differences in the $\theta(t)$ plots. Print graphs of these critical cases and interpret them through annotations in nonmathematical language. Also print phase-plane plots (L versus θ) of these cases.

EXERCISE 7.9[E]   Choose a sufficiently small dt with initial conditions $v_0 = 0$, $\theta_0 = \pi/2$; print graphs showing (over a range of not more than a few degrees) $\theta(t)$ and $L(t)$. Comment on the qualitative comparison of these curves with those you would expect (via equations 7.12) for a mass in free fall; measure dv_y/dt from the $L(t)$ graph to check its value.

EXERCISE 7.10[E]   A further test of the worksheet is to let $g = 0$ and give the pendulum a push (let $v_0 \neq 0$). (You will have to enter dt directly into cell C14 in this case since the formula shown in Screen 7.1 fails when $g = 0$.) Solve equations 7.14 analytically in this case and verify that your worksheet gives the same result.

EXERCISE 7.11[E]   For initial conditions with $v_0 = 0$, the pendulum should oscillate back and forth from its initial position. By inspecting graphs, you can read a rough value of the period of oscillation (time interval from one maximum of θ to another); by inspecting the tabulation in the work block, you can read a more accurate value. Tabulate a list of these values (under the column headings provided on the notes page) and graph these data. (An appropriate graph specification already exists in the distributed PEND worksheet template.) Since the data exist in disparate parts of the worksheet, you may either make penciled notes before moving to the notes page or learn some new spreadsheet commands from the prescription that follows to look at two parts of the worksheet on one screen:

--:   {Home} {ScrollLock}{Right}{ScrollLock} {Right 2}
D1:   /Worksheet Window Vertical
--:   /Worksheet Window Unsynch
       These commands split the screen into two windows that can view different parts of the worksheet.

--:   {Window}{GoTo}{Name}{Enter}
       This introduces {Window}, or {F6}, which can bounce you back and forth between the two parts of the screen and uses {GoTo}, or {F5}, along with {Name}, or {F3}, to get to the work block quickly.

I41:   {ScrollLock}{Down 3}{Right 2}{ScrollLock}
K44:   {Down 2} /Worksheet Titles Horizontal
       The Worksheet Titles command keeps rows of labels from scrolling off screen when you want to scan down the table to look for a return to the initial state after a period of motion.

--:   {Window}
B18:   +$Q_INIT{Enter}
B18:   /Copy B18 to B19..B38
       This prepares the first column of your table for later modification (to list the initial positions you survey).

After the screen is set up this way, you can use the **/Range Value** command to copy the time corresponding to the first return to the initial state to the **period** column on the notes page, then use **/Range Value** to copy the corresponding amplitude to itself (so it won't change during the next recalculation). Use {F6} to move from one window to the other as needed.

7.4 Listings

```
                                                range names:
                                                TITLE2  A1
F6:  "MR^2              G6:  +M*R*R             M       B6      #     I41
F8:  "MgR               G8:  +M*G*R             G       B7      T     K45
                                                R       B8      Q     L45
                                                Q_INIT  C11     WHS   M45
B14:  "dt               C14:  +E14*2*@PI*@SQRT(R/G)  V_INIT C12  LHS   N45
D14:  's      =                                 DT      C14     N     O45
E14:  (T) U 1/20        F14:  '*2*pi*SQRT(R/g)  TITLE1  D2      L     P45
                                                MR^2    G6      E     Q45
                                                MGR     G8
```

```
K44: [W7] ^t      L44: [W7] ^q       M44: [W8] ^whs     N44: ^Lhs
K45: [W7] 0       L45: [W7] +Q_INIT  M45: [W8] +LHS/$MR^2  N45: +L+N*DT/2
K46: [W7] +T+$DT  L46: '?            M45: '?             N46: '?
...
K145..N145
```

```
    O44: ^N   P44: ^L         Q44: ^E
    O45: '?   P45: +M*V_INIT*R  Q45: +L*L/2/$MR^2-$MGR*@COS(Q)+$MGR
    O46: '?   P46: (LHS+N46)/2  Q46: +LHS*N46/2/$MR^2-$MGR*@COS(L46)+$MGR
    ...
    O145..Q145
```

Listing 7.1: The **PEND** worksheet template. To be completed, it needs the essential physics from Schema 7.1 entered in cells marked '? and their copies. Refer to Screens 7.1 and 7.2 for cells that contain only labels or input numbers.

Graph Settings

Name:	QLE_T	*Type:*	XY
Titles:			
1st:	\title1	*2nd:*	\title2
X-axis:	time t [s] -->		

Range		*Format*	*Legend*
X:	K45..K145		
A:	L45..L145	Both	Angle [rad]
B:	P45..P145	Lines	Ang_mom [J s]
C:	Q45..Q145	Symbols	Energy [J]

Graph Settings

Name:	L_Q	*Type:*	XY
Titles:			
1st:	\title1		
2nd:	Phase-plane graph		
X-axis:	angle q [radians] -->		
Y-axis:	angular momentum [J s]		

Range		*Format*
X:	L45..L145	
B:	P45..P145	Both

Graph Settings

Name:	P_QMAX	*Type:*	XY
Titles:			
1st:	\title1		
2nd:	\title2		
X-axis:	amplitude [radians] -->		
Y-axis:	Period [s]		

Range		*Format*
X:	B18..B38	
C:	C18..C38	Lines

Listing 7.2: Graph settings for the PEND worksheet.

Chapter 8

Collisions and Scattering in One Dimension

This chapter contains computer models of two and three particles interacting in one-dimensional motion. Such motions can be observed in gliders on an air track and are of interest because they embody ideas applicable to the interaction of atoms and of elementary particles. In this respect, the air track demonstrations are also models—simplifications that let us explore a limited set of ideas in the hope of gaining insight into deeper ideas. Like all good physical theories, these models succeed in part but have important limitations. The most important limitation here is not the small number of particles that interact or the restriction to one dimension but the omission of quantum ideas essential in the microphysics of atoms and elementary particles. Quantum mechanics does not replace these models but uses and builds upon them, so they are an important step in the right direction.

The analytical tools (from your textbook) applied and illustrated in this chapter are energy and momentum conservation, center-of-mass motion, and—in optional later sections—internal energy and inelastic collisions.

8.1 Two-Body Motions

For two-body interactions, we assume two point particles with masses m_1 and m_2 at positions x_1 and x_2 in one dimension modeled mathematically as the real line $-\infty < x_1, x_2 < +\infty$. We assume a force between them that depends only on their separation

$$x = x_1 - x_2 \tag{8.1}$$

so that a force $F_{x1} = F(x)$ acts on the particle at x_1, and a related force $F_{x2} = -F(x)$ acts on the particle at x_2 in accordance with Newton's third law. There will be a corresponding potential energy $U(x)$ to be included in energy computations to get a conserved energy. Newton's second law gives then

$$m_1 \frac{d^2x_1}{dt^2} = F(x_1 - x_2) \quad , \quad m_2 \frac{d^2x_2}{dt^2} = -F(x_1 - x_2) \quad , \tag{8.2}$$

and the total energy of the two-particle system is

$$E = \frac{1}{2}m_1 \left(\frac{dx_1}{dt}\right)^2 + \frac{1}{2}m_2 \left(\frac{dx_2}{dt}\right)^2 + U(x_1 - x_2) \quad . \tag{8.3}$$

The potential energy $U(x)$ must satisfy

$$F(x) = -\frac{dU}{dx} \tag{8.4}$$

so that this energy be conserved, as Exercise 8.1 confirms. It may seem surprising that this potential energy appears only once in the energy formula although the force $F(x)$ appears twice in Newton's second law (once in each of the two equations 8.2). But when one considers small changes in energy dE that may occur during an interval dt and uses $dx = dx_1 - dx_2$, the contribution from the potential energy $U(x)$ is $dU = (dU/dx)dx = -F(x)dx = -F(x)(dx_1 - dx_2)$ and correctly corresponds to the work done by the two forces $\pm F$ acting on the two different masses. For this reason, we do not associate U with either of these forces alone but refer to it as the **interaction energy** associated with the pair of equal and opposite forces by which each mass acts upon the other.

A consequence of equations 8.2 we want to illustrate in this chapter is momentum conservation and its implications for center-of-mass motion. The sum of the two equations 8.2 is

$$\frac{d^2(m_1 x_1 + m_2 x_2)}{dt^2} = 0 \tag{8.5}$$

since the two F terms cancel. This implies that the center-of-mass location x_{cm}, which is defined by

$$(m_1 + m_2)x_{cm} \equiv (m_1 x_1 + m_2 x_2) \quad , \tag{8.6}$$

changes linearly with time since its first derivative, by equation 8.5, is constant:

$$(m_1 + m_2)\frac{dx_{cm}}{dt} = m_1 \frac{dx_1}{dt} + m_2 \frac{dx_2}{dt} \equiv P_{tot} = \text{const} \quad . \tag{8.7}$$

In the worksheets we construct, the graphs will show how this definition of x_{cm} guides us in organizing the motion. During debugging, the constancy of P_{tot} provides a useful check that we actually implement the calculation we intend.

To write a worksheet to solve equations 8.2, we first rewrite these equations as marching orders to calculate how positions x and momenta $p \equiv mv$ change during each infinitesimal time step:

$$\begin{aligned}
dx_1 &= (p_1/m_1)\,dt \quad , \quad dx_2 = (p_2/m_2)\,dt \quad , \\
dp_1 &= F(x_1 - x_2)\,dt \quad , \quad dp_2 = -F(x_1 - x_2)\,dt \quad .
\end{aligned} \tag{8.8}$$

These equations will be approximated (using the leapfrog or half-step scheme from Section 4.2) by finite-difference equations in the worksheet, and some quantities of interest—such as positions, momenta, and various energies—will be graphed. To simplify writing the worksheet, we will use not SI units but units adapted to the problem at hand. The choice of length and time units will be dictated by the force law we use. For a mass unit, we will use the mass m_1 of one of the particles and then write the second mass as $m_2 = M m_1$ so that M is a dimensionless number, the mass ratio m_2/m_1. By adopting m_1 as our unit, we can use the preceding formulae with the simplification

$$m_1 = 1 \quad , \quad m_2 = M \quad . \tag{8.9}$$

The question, How does our mathematical model behave? always contains an implication that all, or most, possible behaviors are to be surveyed. It is therefore a substantial simplification to reduce the number of parameters varied during the survey. For this reason, we usually want to fix as many parameters as possible by the choice of units. Here we see that we need to include not two independent masses but simply one mass ratio M.

EXERCISE 8.1^{E} Show, by calculating the change in energy dE during an infinitesimal interval dt from equation 8.3, that energy defined this way is conserved as a consequence of Newton's second law as stated in equations 8.2 when the force law and the potential are related by $F(x) = -U'(x) \equiv -dU/dx$. (Hint: collect together all the terms containing $(dx_1/dt)dt = dx_1$ and similarly those containing dx_2.)

8.2 COLLN: Two Interacting Point Particles

The COLLN.wk1 worksheet is to solve Newton's second law from equations 8.8 using the leapfrog method so you should begin building it from a previous example in which this method has been used and debugged. We assume this example is the file CBOND.wk1, constructed in Section 5.2. We will modify it to apply to two particles.

One Mistake at a Time

Numerical computations are full of critical detail, so no one writes a large worksheet without making errors along the way. What is essential is to discover and correct your own errors quickly and to know that they are corrected. An important technique for achieving this goal is to break your work into reasonably sized chunks expected to contain not more than about one error. This avoids the confusion of debugging several interacting errors. For this reason, we will make changes in stages, with tests for proper operation in between.

A reasonable goal for the first set of changes is to produce a correct two-body motion without any interactions. Begin by renaming the worksheet (i.e., save it under a new name)

```
--: /File Save COLLN{Enter}    .
```

We are next going to copy the main computation in the work block, which works for one particle, so the copy will do the same for a second particle. First, inspect the row of initial-value formulae in the work block to see that all constants (such as x_0 and v_0) are marked with **$** signs so that they don't change into something else when copied. Then copy sideways the complete set of four columns containing the labels **x**, **vhs**, **a**, and **v**:

```
L41: /Copy {End}{Right} {End}{Down} to P41
 --: {Calc}
```

After recalculation (**{F9}**), the copied set of four columns should produce exactly the same numbers as the original set. Instead of inspecting the numbers, we can look at a graph:

```
 --: /Graph Name Use X_T{Enter}
/G: A {Escape} P44..P144{Enter} D L44..L144{Enter}
/G: Options Legend A x_2 position{Enter}
/GO: Legend D x_1 position{Enter}
/GO: Format A Symbols D Lines Quit Quit View
 /G: Name Create POSITION{Enter} Name Delete X_T{Enter}
```
This plots the new position column on the same graph with the original position column. With current initial data, the two curves should lie exactly on top of each other.

Revised I/O

To make the computations of x_1 and x_2 independent, they must be supplied with distinct initial data although we will let them continue to use the same equations (solve the same differential equation) until we complete one more check successfully. Return to the home block of the worksheet and rewrite it to correspond to Screen 8.1 which provides places for the initial data for two particles. There are some formulae to be filled in later, as in the area headed **Results**, but the text (labels) can be edited now. To produce this screen, you will need to add a few blank rows for the results section:

```
A12: /Worksheet Insert Row A12..A15{Enter}
```
This command is easily given by pointing; after the keystrokes **/wir**, highlight (in any column) rows 12–15 and press **{Enter}**.

To give effect to the initial data on this screen, they must be connected to the calculations that use them. In setting up these relations it is easiest to throw away all the old cell names and start fresh since there are a lot of unneeded names to delete. Note the locations of the few names you will want retained and reinstate them (or delete all the other names one by one).

```
--: {GoTo} {Name} #{Enter}
```

```
C17: (T) U @PI/30                                                    EDIT
0.1047197551
```

```
         A        B        C        D        E        F        G        H
 1  your name  &  today's date    Case# 1
 2  your class & section          COLLN, two-particle collision in 1 dim
 3                                      Leapfrog integration
 4  --------------------------------------------------------------------
 5  Construction parameters:
 6        Mass ratio  M =              2 = M_2 / M_1
 7        Use one particle's mass M_1 as mass unit; length and time
 8        units are implied by the force law. F_3 = (1/x^5 - 1/x^3)/2
 9  Initial data:
10     M_1:  x1_init       5 L         v1_init     -0.6 L/T
11     M_2:  x2_init       0 L         v2_init      0.1 L/T
12  Results:             init     final    change    check
13                P_1  -0.6000   0.3394   0.9394   0.9333
14                E_1   0.1800   0.0576  -0.1224  -0.1244
15                E_2   0.0100   0.1367   0.1267   0.1244
16  Approximation:
17                dt = @PI/30   T        E var'n =    0.50%
18  --------------------------------------------------------------------
19  Notes page:
20
    COLLN.WK1
```

Screen 8.1: The home screen of the **COLLN** worksheet. Note that the active line here evaluates the formula in cell **C17**. To accomplish this you must first press {Edit}, then {Calc}; to avoid replacing the formula by its value you must press {Escape} before you move from the evaluated cell.

I45: /Range Name Reset
I45: /Range Name Create #{Enter} {Enter}
  It helps to find this location before its name gets temporarily lost.

C17: /Range Name Create dt{Enter} {Enter}
A1: /Range Name Create title2{Enter} {Enter}
D2: /Range Name Create title1{Enter} {Enter}
  These few names need to be restored after a reset wipes out all range names.

B10: /Range Name Labels Right B10..B11{Enter}
E10: /Range Name Labels Right E10..E11{Enter}
  Here we assign enough names to be able to specify independent initial data for the two particles.

C17: @PI/15
C17: /Range Format Text {Enter}
  Since small amplitude oscillations were found to have period 2π in the **CBOND** worksheet, it is convenient to format the *dt* cell so

	K	L	M	N	O	P	Q	R	S
45	time	pos'n	mom_n+1/2	F_on_1	mom_n	----------second particle----------			
46	t	x_1	P1hs	F_x1	P_1	x_2	P2hs	F_x2	P_2
47	0	1	0.01	0	0.01	1	0.003	0	0.003
48	0.314159	1.003141	0.009026	-0.00309	0.009513	1.000942	0.002705	-0.00093	0.002852

Screen 8.2: Some labels in the work screen of the COLLN worksheet at an intermediate step in its construction.

that π can be written visibly.

Now we test the worksheet again. First enter the values $x_1(0) = 1 = x_2(0)$ and $v_1(0) = .01$, $v_2(0) = .003$ in rows 10 and 11. Then bring these values into the work block by means of the formulae in the init data row there.

> **K46:** `/Worksheet Delete Row {Enter}`
> We omit the row of unit reminders.

> **K46:** `/Worksheet Global Label-prefix Center`
> **K46:** *edit labels K46...S46*
> **K46:** `/Range Name Labels Down K46..S46{Enter}`
> The labels over the first formula in each column should be changed to read, in order, t x_1 P1hs...F_x2 P_2, as in Screen 8.2. They name the cells below them, to make writing formulae more understandable.

You can now use these cell names when entering formulae in the init data row of the work block to cause the initial values specified in the I/O block to be used in computation. Schema 8.1 shows these formulae. The force cells need not be changed here, so we continue to use the force law inherited from CBOND with an assumed particle mass $m = 1$ for both particles. Since the initial data are now slightly different, the two graph traces should no longer lie on top of each other but should show simple oscillations with the x_1 trace having a larger amplitude than the x_2 trace.

If some debugging is needed here, the initial data may be changed as desired. You should have two parallel examples of the CBOND motions operating here. These motions of x_1 and x_2 have independent initial conditions but should give identical results when the initial data are the same.

Interacting Particles

Now that we have produced a correctly working model of two particles moving independently of each other, we are ready to change the equations and make the particles exert forces on each other, properly related by action equals reaction. To make the force law easier to change in the future, we will calculate it in just one clearly visible place. This requires a column for $F(x)$ and another where

x_1	P1hs	F_x1	P_1
$x_{1,0}$:	$p_{1,1/2}$:	$F_{1,0}$:	$p_{1,0}$:
$x_1(0)$	$p_{1,0} + F_{1,0}\,dt/2$	$F(x_{1,0})$	$v_1(0)$

x_2	P2hs	F_x2	P_2
$x_{2,0}$:	$p_{2,1/2}$:	$F_{2,0}$:	$p_{2,0}$:
$x_2(0)$	$p_{2,0} + F_{2,0}\,dt/2$	$F(x_{2,0})$	$v_2(0)$

Schema 8.1: Initial data for the COLLN worksheet at an intermediate step in its construction where the two particles move independently of each other.

we can calculate (and later plot) the relative distance $x = x_1 - x_2$. Then the force columns F_x1 and F_x2 are changed to yield $\pm F$ so that the computed motion actually uses the desired force.

P46: /Worksheet Insert Column {Right} {Enter}
P46: F
Q46: x
P46: /Range Name Labels Down {Right} {Enter}
This creates two empty columns and gives them labels.

Q47: +x_1-x_2
P47: (x^-5-x^-3)/2
P47: /Copy P47..Q47 *to* P48..P147
The same force law $F(x) = (x^{-5} - x^{-3})/2$ that was used in the CBOND worksheet will be used here.

N47: +F
N47: /Copy N47 *to* N48..N147
T47: -F
T47: /Copy T47 *to* T48..T147
--: {Calc}
This is the step where the force is brought to bear on the motion of the particles. In accordance with action equals reaction, equal but opposite forces are applied to the two particles.

While copying these new formulae down the columns, you may have seen the value ERR displayed occasionally. This is of no concern since many of the numbers going into the calculations are left over from a previous—and now irrelevant—calculation. If the initial data and time step are chosen reasonably, the calculation should give reasonable answers after {Calc} (i.e., {F9}) is pressed. Try the data shown in Screen 8.1.

x_1	P1hs	F_x1	P_1
$x_{1,0}$: $x_1(0)$	$p_{1,1/2}$: $p_{1,0} + F_{1,0}\,\mathbf{dt}/2$	$F_{1,0}$: $+F_0$	$p_{1,0}$: $m_1 v_1(0)$
$x_{1,1}$: $x_{1,0} + p_{1,1/2}\mathbf{dt/m_1}$	$p_{1,3/2}$: $p_{1,1/2} + F_{1,1}\,\mathbf{dt}$	$F_{1,1}$: $+F_1$	$p_{1,1}$: $(p_{1,1/2} + p_{1,3/2})/2$

x_2	P2hs	F_x2	P_2
$x_{2,0}$: $x_2(0)$	$p_{2,1/2}$: $p_{2,0} + F_{2,0}\,\mathbf{dt}/2$	$F_{2,0}$: $-F_0$	$p_{2,0}$: $m_2 v_2(0)$
$x_{2,1}$: $x_{2,0} + p_{2,1/2}\mathbf{dt/m_2}$	$p_{2,3/2}$: $p_{2,1/2} + F_{2,1}\,\mathbf{dt}$	$F_{2,1}$: $-F_1$	$p_{2,1}$: $(p_{2,1/2} + p_{2,3/2})/2$

Schema 8.2: The interaction of two particles through a force obeying Newton's third law in the COLLN worksheet implements equations 8.8.

Validation

How do we now check that the worksheet is doing the calculation we want done? The previous modifications of the CBOND worksheet were all intended to leave working calculations identical to those we had seen in the unmodified worksheet. Now, however, the physics has changed by making the force on each particle depend on the position of the other.

Checkpoints What are the simple features of motion we can figure out without a computer and thus use to check that we haven't overlooked something or made typographical errors? One is equilibrium. The separation x can stay constant when the interaction force $F(x) = (x^{-5} - x^{-3})/2$ is zero, which occurs when $x \equiv x_1 - x_2 = 1$. Thus we can choose initial conditions with $x_1 - x_2 = 1$ and $v_1 = 0 = v_2$ and check that $x_1(t) - x_2(t) = 1$ continues to hold for later times; the same should be true for initial conditions with $x_1 - x_2 = 1$ and any initial velocity, provided $v_1 = v_2$. A second feature that can be predicted without a computer is that the center of mass will remain at rest if it starts at rest. Thus all motions that begin with $v_1 = -v_2$ should preserve this condition. (The worksheet at this point has $m_1 = m_2$.) A third feature is that the center of mass moves at constant velocity; a fourth, that the total momentum is constant; a fifth, that energy is conserved. These last three are not easy to verify just by inspecting the POSITION graph. Instead we need to enlist the aid of the computer to compute and graph the features we want to verify. Since all these quantities involve the particle masses, let us edit the equations to include two different masses, $m_1 = 1$ and $m_2 = M m_1 = M$, as indicated in Schema 8.2, and then proceed to write the additional columns of data:

V46: *edit labels V45...AB46*

	I	J	K	L	M	N	O	P
45	name:		time	pos'n	mom_n+1/2	F_on_1	mom_n	Force
46	labels:		t	x_1	P1hs	F_x1	P_1	F
47	init data:		0	5	-0.60020	-0.00384	-0.6	-0.00384
48	typical row:		0.104719	4.937147	-0.60062	-0.00400	-0.60041	-0.00400
49	copied rows:		0.209439	4.874250	-0.60105	-0.00418	-0.60084	-0.00418

	Q	R	S	T	U	V	W	X
45	sep'n	----------second particle----------				C of Mass	tot mom	vrel
46	x	x_2	P2hs	F_x2	P_2	cm	P	Vr
47	5	0	0.200201	0.00384	0.2	1.666666	-0.4	-0.7
48	4.926664	0.010482	0.200620	0.004009	0.200410	1.652704	-0.4	-0.70061
49	4.853263	0.020986	0.201059	0.004188	0.200840	1.638741	-0.4	-0.70126

	Y	Z	AA	AB	AC	AD	AE	AF
45	--------------Energies--------------							
46	E_1	E_2	E_int	E_tot				
47	0.18	0.01	-0.0098	0.1802				
48	0.180246	0.010041	-0.01008	0.180200				
49	0.180504	0.010084	-0.01038	0.180200				

Screen 8.3: The top rows from the work block of the COLLN worksheet.

V46: /Range Names Labels Down V46..AB46{Enter}
Follow Screen 8.3 and produce column headings for a variety of data for graphing and analysis. Use the label row to give names to the first computed cell in each column.

The formulae to be entered in columns V..AB are given in Schema 8.3, but with mass units as agreed from equation 8.9. Recall that a schema gives mathematical formulae. You must assign cell names and translate these formulae into spreadsheet syntax.

Graphs Let us now graph these data: The center-of-mass position x_{cm} can be plotted on the graph named POSITION, which we made on page 106. (Plot it as color C, formatted simply as Lines, and do not forget that the /Graph Name Create POSITION command must be repeated so that these changes will be remembered with the previous specifications for this named graph.) Another informative graph, called MOMENTUM, can graph the three momenta p_1, p_2, and P_{tot}. The total momentum should be a constant. An ENERGY graph (Figure 8.1) can then be designed to plot the kinetic energies E_1 and E_2 of the two particles, their interaction energy U, and the total energy E_{tot}, all on one graph. If the force law and the potential energy are consistent with each other, and if the time step is sufficiently small, the total energy should remain nearly constant. (In a scattering event, the interaction energy should be small at both the beginning and the end of the calculated motion when the two particles are far from each other.)

These calculations complete the diagnostic tools that analytical arguments let us build into the worksheet. They also illustrate important concepts in

labels:	cm	P	Vr
init	x_{cm}:	P_{tot}:	v_{rel}:
data:	$\dfrac{m_1 x_1 + m_2 x_2}{m_1 + m_2}$	$p_1 + p_2$	$\dfrac{p_1}{m_1} - \dfrac{p_2}{m_2}$

E_1	E_2	E_int	E_tot
E_1:	E_2:	U:	E_{tot}:
$\dfrac{p_1^2}{2m_1}$	$\dfrac{p_2^2}{2m_2}$	$\dfrac{1}{8x^4} - \dfrac{1}{4x^2}$	$E_1 + E_2 + U$

Schema 8.3: The formulae to be entered in columns V..AB in the work block of the COLLN worksheet are given here. They should be copied down through row 147.

mechanics: equilibrium, center-of-mass motion, momentum conservation, and energy conservation.

Completion

To complete the construction of the worksheet we will add a phase-plane graph and a small output table. Since there are two coordinates here, plus combinations of them such as the relative displacement and the center of mass, as well as corresponding momenta, we could define several phase planes. The most interesting graph from this assortment plots the relative velocity $v_{rel} = v_1 - v_2 = p_1/m_1 - p_2/m_2$ of the two particles as a function of their relative displacement $x = x_1 - x_2$. Construct this graph and name it PHASE_PLANE. In the next section we will study the reasons why this graph is so similar to that of the single-particle motion studied in Chapter 4. Another useful graph can be called XVF_T; it plots the relative position $x = x_1 - x_2$, the relative velocity $v_{rel} = v_1 - v_2$, and the interaction force $F(x)$ as functions of time.

The output table already labeled on the home screen (Screen 8.1) reports beginning-to-end changes of the kinetic energies of the two particles and of one of their momenta. The following listing shows how most of the table is computed:

```
A12: U ^Results:
        C12: U ^init        D12: U ^final       E12: U ^change
B13: U "P_1
        C13: (F4) +$P_1     D13: (F4) +$O$147   E13: (F4) +D13-C13
B14: U "E_1
        C14: (F4) +$E_1     D14: (F4) +$Y$147   E14: (F4) +D14-C14
B15: U "E_2
        C15: (F4) +$E_2     D15: (F4) +$Z$147   E15: (F4) +D15-C15
```

The (F4)s here show that these cells have been formatted to display numbers

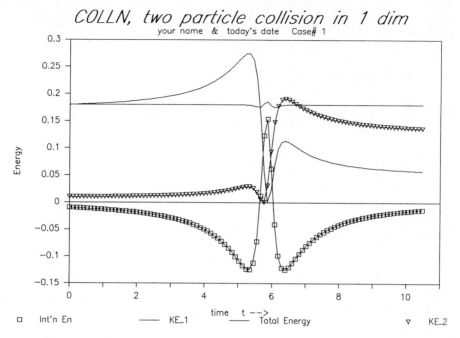

Figure 8.1: Energies during the collision corresponding to the data shown on Screen 8.1.

in a fixed-point format with four digits following the decimal point. To install this format, use the command:

C13: /Range Format Fixed 4{Enter} C13..F15{Enter}

The formulae in the **check** column of these results will be added in Exercise 8.10 later. Another convenience that may be added to the I/O page is a report on the constancy of the energy (to flag cases where dt can be seen to be too large by this criterion):

F17: @STD(AB47..AB147)/(@AVG(Y47..Y147)+@AVG(Z47..Z147))
F17: /Range Format Percentage 2{Enter} {Enter}

The formula here uses the **@STD** function, which computes the standard deviation of a range of numbers, i.e., the root-mean-square deviation of these numbers from their average. It is here reported as a percentage of the average kinetic energy, which is chosen as a readily available typical energy in the computation. Unless you enlarge the worksheet with many more rows to allow smaller time steps, you will not be able to run a complete collision with much better than 1% energy nonconstancy.

EXERCISE 8.2^C Complete the COLLN worksheet as described in the prescriptions, screens, and schemata above. This can be done most quickly if the quality-assurance tests are done as suggested at intermediate stages in the worksheet construction. Print out a copy of the home screen and of a screen showing the first rows of columns K through R in the work block, with calculations corresponding to the data shown in Screen 8.1. Also print the formulae used in your init data and typical rows, which are range K47..AB48 in Screen 8.3.

EXERCISE 8.3^C In the COLLN worksheet, define named graphs ENERGY, MOMENTUM, POSITION, PHASE_PLANE, and XVF_T as described previously, and submit {PrintScreen} examples of them using the data in Screen 8.1 as input. Be sure each graph has appropriate legends and axis titles, with distinctive formats for the different quantities plotted, so that the information content is clear. (If the expected behaviors of the center of mass, total momentum, and total energy are not found, recheck your worksheet for mistakes.)

8.3   Exploration

Now that you have a working mathematical formulation of two particles interacting via a force $F(x_1 - x_2)$, you should use it to become familiar with the motions it allows. One example of a collision between the two particles arises from the sample data in Screen 8.1, which you may have used while debugging the COLLN worksheet. The graphs show intermediate stages of the interaction (while the force is significant) and allow you to visualize the predicted motion in detail, an advantage this worksheet has over a simple analytical calculation that predicts, for an elastic collision, final velocities in terms of initial velocities. In particular, if you think of this force law as a model for a chemical force between two atoms, notice that the atoms never stick together when they bump into each other—they simply rebound after colliding.

Bound States

This worksheet does, however, give a better picture of an idealized two-atom molecule than did our CBOND worksheet—it shows both atoms and allows them to have different masses. To run this worksheet as a molecular model, set up initial conditions where the two particles start at their equilibrium separation $x = 1$ and let one of them (or both) be moving. If the initial velocities are not too different, the particles should stay bound to each other (x should oscillate between finite limits) while the center of mass moves at constant velocity. We thus have a mathematical model that allows two particles to *be* bound but not to *become* bound.

Equal-Mass Collisions

Elastic collisions between equal-mass particles have an especially simple behavior. You should be able to prove just from the conservation of energy and momentum in the collision that the net result is for the two particles to exchange their momenta. To see this in the numerical model of a collision, retrieve your saved worksheet **COLLN** with the input data shown in Screen 8.1, and change the mass ratio to $M = 1$. Then inspect the graphs that result in both this case (call it 3a) and two others: (3b) keep the same relative velocity but set $v_2(0) = 0$, and (3c) keep the same relative velocity but make the total momentum zero.

Galilean Relativity

The preceding example provides an illustration of Galilean **relativity**: many important features of the motion are unchanged when the initial velocities of all objects are changed by the same amount. The three cases just specified above satisfy this condition. The **PHASE_PLANE** graph, which plots only relative quantities, should be exactly the same in all three cases. Another relativistic invariant you might notice is the momentum transfer, i.e., the amount of momentum one particle gives up to the other. The most important relativistic invariant here is the equations of motion themselves: equations 8.2 remain exactly the same (require no modifications to the work block) if one decides to describe the motion by reference to an observer moving with constant velocity V_{obs}. This change requires the replacements

$$x_i \rightarrow x_i - V_{obs}t \quad , \quad i = 1, 2 \tag{8.10}$$

but the additional term cancels in $x = x_1 - x_2$ and does not survive differentiation in the acceleration terms of equations 8.2.

Global Conservation Laws

The numerical model of a collision allows us to study energy and momentum on a millisecond-by-millisecond basis, but the experiments that lead us to believe that our numerical models should incorporate energy and momentum conservation compare more widely separated beginning and end states. To illustrate this aspect of the conservation laws, you can modify the force law (without at first making the energy formula consistent with it) and see that energy and momentum are still conserved on the global scale of a complete collision even if you do not have a good definition of energy during the interaction. Change the force law (column P) to read $F(x) = x^{-5}/2$ and rerun some set of initial conditions for which you have printouts of the I/O block under the previous force law. You should find that the momentum and energy transfers calculated in the **results** area are nearly the same, independent of the force law used. (They would be precisely equal only if the initial and final separations x were large enough to make the interaction energy unnoticeable at the beginning and end in both cases.) The energy graph will not show the total energy being conserved, however, if you have not changed the formula for the interaction energy. You can change the interaction energy to zero to suggest ignorance of how to calculate it, with similar results. But note that the energy changes ΔE_1 and ΔE_2 are (approximately) equal and opposite whenever the collision computation begins and ends with the two particles far apart. If you modify the interaction-energy $U(x)$ column to compute the potential energy corresponding to the force you are using, the energy graphs should again show a nearly constant total energy. You can use many other force laws and get the same results. The only requirements are that the force not allow the two particles to pass through each other and that it approach zero fairly rapidly as the separation gets large. A repulsive force law, such as $F(x) = x^{-n}$ with n large, will require using a small time step dt since the force changes rapidly from small to large with the entire interaction taking place over a short distance in a short time.

Analytical Checks

In one-dimensional elastic collisions, energy and momentum conservation allow one to predict the final momenta of the two particles from the initial momenta by using simple algebraic relationships. You should derive these formulae and program them into the **check** column (in the **Results** area of the I/O block) of your worksheet, then use them to compare with the numerically computed momentum transfer and energy exchanges reported in the **change** column. The algebra is simplest if you consider the special case where one mass, say m_2, is initially at rest. (These calculations are not applicable when the interaction energy U cannot be neglected at the beginning and end of the motion, e.g., in bound states.)

Limiting Cases

It is often instructive to study limiting cases, those in which some parameter is very large or very small. Usually some simplification occurs, but in numerical work it is also possible that difficulties arise. Here the only (dimensionless) construction parameter is the mass ratio $M = m_2/m_1$. One extreme case to try is large M, say $M = 10^8$. In this case, the center of mass essentially coincides with x_2, the position of the large mass. You will find that the energy graphs are overwhelmed by the kinetic energy of m_2 unless you choose $v_2(0) \simeq 0$. With $v_2(0) = 0$, the motion of m_1 should be indistinguishable from the motions programmed in **CBOND**, even down to numerical detail. For instance, in a bound state with a very small amplitude vibration (near the minimum possible total energy), the vibration should have a period (repetition time) of 2π in the units we have chosen.

Surprisingly, the results are not so simple if we make m_1 the heavy mass. Try $M = 10^{-8}$ and set $v_1(0) = 0$. The results are numerically different from the previous case. In addition, they may be difficult to obtain because you must explore to find an appropriate step size dt. The only apparent difference between the cases $M = 10^8$ and $M = 10^{-8}$ is that in one the small mass and in the other the large mass is called m_1. They could have been the same two masses, differently labeled. How can our mathematical model be influenced by what we choose to call m_1? It is so influenced because we let m_1 be the mass unit translating physical magnitudes (e.g., a force of $0.3\,\mathrm{N} = 0.3\,\mathrm{kg}\,\mathrm{m}^2/\mathrm{s}^2$) into pure numbers stored in our worksheet. The next section explains a more convenient way to choose the mass unit. The choice there is called the reduced mass μ and is defined to be completely symmetric between the two particles:

$$\mu \equiv \frac{m_1 m_2}{m_1 + m_2} \quad . \tag{8.11}$$

With this definition, μ becomes (nearly) the smaller mass whenever the masses are very different.

EXERCISE 8.4$^\mathrm{C}$ Try a variety of initial conditions that represent the two particles approaching each other (i.e., with nonzero relative velocity) beginning at a separation $x \gg 1$ in order to see (by this numerical experiment) that the computed motion never suggests they will end up bound together at separations much closer than their initial distance (i.e., two free atoms with this single-particle structure do not spontaneously form a molecule).

EXERCISE 8.5$^\mathrm{H}$ Prove analytically from energy-momentum conservation that, as in the previous exercise, capture will not occur for any other force law $F(x)$ for which the corresponding potential energy $U(x)$ approaches zero when the separation x is large. (Hint: some energy is tied up in the center-of-mass motion; see Section 8.4 and Exercise 8.18.)

EXERCISE 8.6$^\mathrm{C}$ Choose initial conditions that show a bound state (vibration where x stays near 1) with the center of mass moving. Mark this by a notation such as Case #2 added to the information in the title2 cell (A1) so that it will appear on your printouts. Then make printed copies of your I/O block or home screen and of all four graphs. Be sure your worksheet has been recalculated with the desired input data before making any printouts.

EXERCISE 8.7$^\mathrm{P}$ Produce printouts of the I/O block and of the POSITION and the PHASE_PLANE graphs for the three cases described under the Equal-Mass Collisions heading, using a note in the title2 cell to associate the data with the resulting graphs. Point out some quantities that are not relativistic invariants.

EXERCISE 8.8$^\mathrm{P}$ Change the mass ratio back to 2 or more and repeat the previous exercise (but label the cases 4a, b, c) to verify that the relativity principle holds in the general case ($M \neq 1$). Check the invariance of the phase-plane graphs and momentum transfer.

EXERCISE 8.9$^\mathrm{P}$ Change the force law as suggested under Global-Conservation Laws and print out I/O screens (labeled as case 5).

EXERCISE 8.10$^\mathrm{E}$ Assume that U can be neglected at the beginning and end of a collision (when $x \gg 1$) and consider a case where one particle is initially at rest $v_2(0) = 0$. From conservation of energy and momentum, derive formulae for the final values of v_1 and v_2 after the collision. From these, calculate analytically the changes your worksheet reports for p_1, E_1, and E_2, and enter these formulae into the check column on the I/O page of the COLLN worksheet. Provide your derivation of the formulae as well as printouts of two different cases of collisions where the numerical changes nearly agree with the analytical predictions.

EXERCISE 8.11^{H} Repeat the previous exercise without the restriction $v_2(0) = 0$. A somewhat tricky first step in the algebra is to prove that the relative velocity simply reverses in the collision (as in all the phase-plane plots). If velocities before and after the collision are called v_i and u_i respectively, this reversal can be restated as $v_1 + u_1 = v_2 + u_2$ and derived as the quotient of the energy-and-momentum-conservation equations after they are suitably arranged.

EXERCISE 8.12^{E} Set $M = 10^8$ and $v_2(0) = 0$ and find initial conditions that produce small oscillations (changes in x much less than the equilibrium separation). Print a graph and corresponding I/O page showing that the oscillation period is very nearly 2π. (Choose a graph that shows the oscillations clearly, either by the choice of variable to plot or by manual scaling to zoom in on the variations.) Print a second example showing the same period with oscillations still smaller by a factor of at least ten.

EXERCISE 8.13^{E} Set $M = 10^{-8}$ and $v_1(0) = 0$ and find initial conditions that produce small oscillations. Print a graph and corresponding I/O page from which you report an oscillation period very different from 2π. Print a second example with $M = 1$ and find the period in this case (again keeping the amplitude small).

8.4 ♠ Reduced Mass

Equivalent Single-Particle Motion

The physics of two interacting masses can be divided into two independent parts: the motion of the center of mass of the pair and their interaction. This section studies the mathematical details that establish this in a precise way. You have already seen indications of this split. The motion of the center of mass is extremely simple: it moves with constant velocity in the absence of external forces. The phase-plane plot of $v_1 - v_2$ against $x_1 - x_2$ showed relative motions uninfluenced by the center-of-mass motion, as did the graph (XVF_T) of these two variables against time. We therefore look for the equations that govern the motion of the relative coordinate $x = x_1 - x_2$. From

$$\frac{d^2x_1}{dt^2} = \frac{1}{m_1}F(x_1 - x_2) \quad , \quad \frac{d^2x_2}{dt^2} = -\frac{1}{m_2}F(x_1 - x_2) \qquad (8.12)$$

we find, by taking the difference of these two equations, that

$$\frac{d^2(x_1 - x_2)}{dt^2} = \left(\frac{1}{m_1} + \frac{1}{m_2}\right)F(x_1 - x_2) \quad . \qquad (8.13)$$

This motivates the definition

$$\frac{1}{\mu} \equiv \frac{1}{m_1} + \frac{1}{m_2} \qquad (8.14)$$

which is the same as equation 8.11. With this definition (and $x_1 - x_2 \equiv x$), the preceding differential equation can be rewritten as

$$\mu\frac{d^2x}{dt^2} = F(x) \qquad (8.15)$$

which has the same form as the equation we studied in the CBOND worksheet. That is, the relative motion of two interacting particles satisfies the same differential equation as a single particle interacting with an immovable center of force. The graphs drawn to display the solution of the one-body problem also display the solution of the two-body problem. All we need to do is replace the position x in the single-body problem with the relative position $x = x_1 - x_2$ in the two-body problem and replace the single-body mass m with the reduced mass μ.

The Internal Time Scale

With the computer's assistance solving a two body problem is nearly as easy as solving a one body problem—we need only copy and edit the one solution to produce the second solution. Also, by explicitly treating two particles we are able to draw informative graphs showing both of their motions. What is most useful numerically from the reduced mass concept is insight into the natural time scale for the motion. If the one-particle problem had a certain time scale when its mass m was taken as the mass unit, then the two-particle problem will have the same time scale when the reduced mass μ is taken as the mass unit. We can therefore edit the COLLN worksheet to use this mass unit. It will then run with the same time step whether the mass ratio m_2/m_1 is 10^8 or 1 or 10^{-8} and thus make exploration easier and more understandable. Also, since equation 8.15 does not contain any mass except μ, we learn that varying the mass ratio during our exploration will show nothing new in the relative motion that cannot be absorbed into the choice of mass unit μ.

Improving the Worksheet

To change from using m_1 as the mass unit in the computation in the COLLN worksheet to using μ, you should first edit the I/O page to look like Screen 8.4. Insert a new row 7 to contain the values of the individual particle masses m_1 and m_2 computed in terms of the mass ratio $M = m_2/m_1$. The formulas needed are

$$m_1 = \mu(1 + 1/M) \quad , \quad m_2 = \mu(1 + M) \tag{8.16}$$

but with $\mu = 1$. Then name the cells displaying these values: cell C7 becomes M_1, and F7 becomes M_2. You must then edit the first two computational rows in the work block so that they use the values \$M_1 and \$M_2 for m_1 and m_2 instead of the previously used numerical values 1 and \$M. Follow Schemata 8.2 and 8.3 and be sure to copy corrected formulae from the typical line down the full length of the columns.

EXERCISE 8.14^C Edit the COLLN worksheet as described above to use $\mu = 1$ as the unit of mass. Print out the home screen and a listing of formulae in row 7 and the initial and typical rows in the work block, now rows 48 and 49.

```
C7: 1+1/M                                                          READY
```

```
         A        B        C        D        E        F        G        H
 1  your name  &  today's date
 2  your class & section        COLLN-mu, two-particle motion in 1 dim
 3                                    Leapfrog integration
 4  -----------------------------------------------------------------------
 5  Construction parameters:
 6          Mass ratio  M =            2 = M_2 / M_1
 7             M_1 =      1.5 mu          M_2 =        3 mu
 8          Use the reduced mass mu as mass unit; length and time
 9          units are implied by the force law. F_3 = (1/x^5 - 1/x^3)/2
10  Initial data:
11     M_1:   x1_init        7 L          v1_init     -0.4 L/T
12     M_2:   x2_init        0 L          v2_init      0.6 L/T
13  Results:            init    final   change   check
14               P_1  -0.6000  1.4025  2.0025   2.0000
15               E_1   0.1200  0.6557  0.5357   0.5333
16               E_2   0.5400  0.0068 -0.5332  -0.5333
17  Approximation:
18            dt = @PI/25  T        E var'n =    0.67%
19  -----------------------------------------------------------------------
20  Notes page:
    COLLN-MU.WK1
```

Screen 8.4: The I/O page of COLLN modified to use μ as the mass unit.

EXERCISE 8.15^C Print a sequence of graphs with corresponding I/O pages show-ing small oscillation for a variety of mass ratios M as in Exercises 8.12 and 8.13, but use the revised worksheet with μ the unit of mass.

EXERCISE 8.16^E Print a sequence of graphs and I/O pages showing collisions all with the same initial positions, velocities, and dt values, but for a range of mass ratios M such as 0.01, 0.5, 1, 2, 100. Use the revised worksheet developed in this section. (Also, verify that the phase-plane graph is identical in every case, but don't print these graphs.)

EXERCISE 8.17^E Confirm that equations 8.16 are correct by showing that they correctly lead to M and μ when substituted into the definitions $M = m_2/m_1$ and $\mu = m_1 m_2/(m_1 + m_2)$.

EXERCISE 8.18^H Show that equation 8.3 can be rewritten as

$$E = \frac{P_{tot}^2}{2M_{tot}} + \frac{p^2}{2\mu} + U(x) \tag{8.17}$$

where $p \equiv \mu v_{rel} = \mu dx/dt$. The first term can be called the energy of the center-of-mass E_{cm}; the remaining two terms can be called the internal energy E_{intern} of the two-particle system. Show that the internal energy is, by itself, conserved.

8.5   ♠ SCAT1: Scattering in One Dimension

In nuclear and in high-energy particle physics, most of the experimental information about subatomic forces is obtained from scattering experiments. In such experiments, two particles are made to collide by throwing one at the other at known energies; the results are then measured when they (or other debris) fly apart again. This approach is used since no laboratory apparatus can hold on to a proton and a neutron and slowly push one against the other to measure directly the forces they exert on each other when they are close enough (10^{-15}m) for nuclear forces to be strong. The aim of this section is to illustrate that observations made at a great distance from the interacting particles can yield results that shed light on the shorter range forces between them. (Because we work here with motion in only one dimension, the types of observations considered are a bit different from those customary in three-dimensional microphysics, where scattering angles are easily measured but calculations are more complicated.)

We will use the F_4 force as a model. It could be realized (in some approximation) by setting up magnets on an air track as suggested in Figure 8.2, although the magnet pair between which the third magnet passes would be in

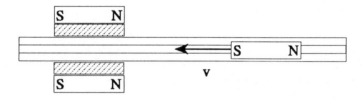

Figure 8.2: Magnetic forces with an air track.

a fixed position in a practical apparatus. You should first verify (Exercise 8.19) that, corresponding to force law F_4 from equation 5.6 (page 72), the potential energy is

$$U_4(x) = \frac{5}{72}(1 - 6x^2)\left(\frac{5}{2 + 3x^2}\right)^2 \quad . \qquad (8.18)$$

This is a rather complicated expression, so you should graph it to see its main features. You can do this by modifying the graph template from Screen D.3 in Appendix D on page 216. These features include minima at $x = \pm 1$ where $F_4(\pm 1) = 0$ and a maximum at $x = 0$ where, necessarily, $F_4(0) = 0$. Newton's second law, $\mu\, d^2x/dt^2 = -dU/dx$, will accelerate the particle separation toward the potential minima and away from the maximum. Using that as a guideline, you can see that this shape of potential qualitatively fits the forces you can imagine, or experimentally feel, between magnets arranged as in Figure 8.2. (If this apparatus is explored in quantitative detail as a long-term student project, it would be appropriate to use a more general force law, as suggested in Exercise 8.24, and to adjust the parameters of that law to fit the observations.)

Forward and Backward Scattering

Worksheet SCAT1 is provided in the student software package; it calculates the motion of two particles interacting via the force law $F_4(x)$ using a format similar to the COLLN worksheet you wrote earlier in this chapter. As a first exploration or numerical experiment using this theoretical model, let us treat particle 2 as the scattering target and particle 1 as the incident probe. Thus, choose initial conditions with $x_{2,init} = 0 = v_{2,init}$ and explore various choices of $v_{1,init}$. You should find two possible cases: forward and backward scattering, i.e., scattering through angles of, respectively, 0 degrees and 180 degrees. The incident particle either continues after interaction in its original direction (forward) or rebounds toward its initial position (backward). What features of the potential energy U_4 control the critical energy of the incident particle that divides the forward from the backward scattering result (Exercise 8.20)? In experiments for which these calculations are a simplified model, there will also be a third case, capture, where the incident particle becomes bound to the target. What simplifications in the model prevent this outcome from the present calculations?)

Time Delay

How do physicists see inside the proton to find quarks there? It takes more than one number to suggest a complete picture of microscopic entities below the resolution of conventional microscopes. The potential $U(x)$ is our representation of invisible internal structure in the present example, analogous to quarks in current research, and we now seek more data that can be measured far from the target but bears on the shape of $U(x)$. In forward scattering, the time required by the incident particle to pass through the target is such a datum. To make the data independent of details like the distance to the accelerator mechanism that provides the incident particle with its initial velocity, we define the time delay as the difference between the time taken by the interacting particle and the time that would be required were there no interaction. Thus we define $T_{delay} = (t_{final} - t_{initial})_{actual} - (t_{final} - t_{initial})_{const\ vel}$ or

$$T_{delay} = (t_{final} - t_{initial}) - (x_{final} - x_{initial})/v_{init} \qquad (8.19)$$

where x is the relative displacement of the two particles, and v is their relative velocity. This quantity can be calculated and graphed by the SCAT1 worksheet; it is normally graphed as a function of energy. If a comparable set of experimental measurements were available, it would be a severe constraint on a model potential energy function $U(x)$ used in computations to reproduce the observed data.

For a very simple example where the time delay data can be related to the interaction potential $U(x)$, consider the two different potentials $U_+(x)$ and $U_-(x)$ defined by

$$U_\pm(x) = \begin{cases} 0 & \text{if } |x| > a/2 \\ \pm\mu v_c^2/2 & \text{if } |x| \le a/2 \end{cases} \tag{8.20}$$

Here v_c and a are two constants that determine the height and width of this square-shaped potential. From analytical arguments (Exercise 8.22: energy conservation yields the (constant) relative velocity when the particles are near each other in terms of the initial relative velocity) one can see that these two constants can be obtained by observations only at low and high initial velocities. Details of the shape of the graph of T_{delay} against energy at intermediate energies would distinguish between potentials of this simple square shape and others, such as U_4.

EXERCISE 8.19^H Verify, by differentiating equation 8.18, that the force and potential are correctly related by $F_4 = -dU_4/dx$.

EXERCISE 8.20^C By numerical experiments using the SCAT1 worksheet as suggested in previous text, find examples of both forward and backward scattering and locate the initial energy that separates these cases. List the critical values of both $m_1 v_1^2/2$ and $\mu v^2/2$ for several values of $M = m_2/m_1$ and see if you can find a rule that fixes the critical energy. Study the ENERGY graph for different cases and see if you can relate the critical energy to some aspect of the potential energy function U_4.

EXERCISE 8.21^C Print the time-delay graph from the SCAT1 worksheet and verify two of the data points by finding initial conditions (and a time step choice) that give approximately the same results. Circle the points you are checking on your graph and print the I/O screen showing each datum you verify.

EXERCISE 8.22^H For the potential $U_-(x)$ from equation 8.20, find the relative velocity v when $|x| < a/2$ in terms of the initial relative velocity v_i. Then find a formula that gives the time delay for any initial velocity v_i. Show that the separate values of the two parameters a and v_c can be extracted from a graph of T_{delay} as function of v_i by finding approximate limiting forms for this function at low and high v_i that each depend on a different combination of the two constants.

EXERCISE 8.23^H Provide an analysis similar to the preceding exercise, but for the potential $U_+(x)$.

EXERCISE 8.24^H A generalization of the F_4 force law that has a qualitatively similar behavior but is different in numerical detail is

$$F_{4'}(x) = \frac{x}{2}(1-x^2)\left(\frac{s^2+1}{s^2+x^2}\right)^{(n+3)/2} . \tag{8.21}$$

Verify that it satisfies $F(\pm 1) = 0$, which sets the unit of length, and $F'(\pm 1) = -1$, which sets the unit of time (given $\mu = 1$). Note that $F(x) \sim -x^{-n}(s^2+1)^{(n+3)/2}/2$ for large x. What values of s and n make this formula yield the force law F_4?

EXERCISE 8.25^{H} Define a potential $U(x)$ from the force $F(x)$ by

$$U(x) = \int_{x}^{\infty} F(x)\,dx \quad . \tag{8.22}$$

Show that this potential satisfies $F = -dU/dx$. Also evaluate this integral using the force from the preceding exercise to obtain an algebraic formula for $U_{4'}(x)$ when $n > 1$. (Hint: the substitution $z = s^2 + x^2$ simplifies the integration.)

8.6 ♠ INELC: Inelastic Collisions

An inelastic collision conserves momentum but not energy. Yet we expect energy to be conserved in *all* interactions on the microscopic scale. In this section you will explore a simple model that shows how the two preceding statements fit together. The two statements differ in the way we think of energy. In an inelastic collision we keep account of only the more visible mechanical energy—the energy associated with the macroscopic motion of the center of mass motion of each body—but ignore internal energies such as thermal energy (e.g., random motions of the microscopic molecules) and internal vibration energy (sound waves).

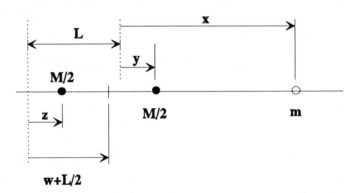

Figure 8.3: Inelastic collision model. The target is a body of mass M composed of two point particles, each of mass $M/2$, located at positions y and z, and held together by forces $F_{\text{int}} = \pm(M/4)\omega_0{}^2(y - z)$. The projectile is a simple point particle of mass m that interacts with the target through forces $F_{ext} = \pm(M/2)\omega_0^2\ell[\ell/(x-y)]^{n+1}$ applied to m and to the particle at y. The center of mass of the target body is located by the coordinate $w = (y + z)/2$. The equilibrium separation L of particles y and z does not appear in the force laws since the base points for these two coordinates were chosen to have that same separation.

Modeling Internal Structure

The model here allows a distinction between total energy and macroscopic mechanical energy by letting one of the two bodies in the collision be compound. To make the compound body as simple as possible, it is made up of only two particles (compared to the 10^{23} atoms in a gram of aluminum). By doing numerical experiments with this model you explore examples in which the action of conservative forces allows macroscopic energy to be converted (under some conditions) into internal energy. Three point particles move along a single one-dimensional line as shown in Figure 8.3. The positions of the masses are called x, y, and z to avoid using a lot of subscripts in the notation. One body, which we call the projectile, has mass m and coordinate x. It has no internal structure. The second body, or target, in this possibly inelastic collision is compound and is made up of two "atoms" at y and z. The total mass of the target will be M, and its center of mass is described by the coordinate $w = (y+z)/2$; this formula presumes then that the two particles making up M have equal masses $M/2$. We next need to make some simplified model of the forces, both internal and external, that act here. We will choose one force to hold the two parts of the target together and a second force to make the target reflect the projectile. So that energy is conserved at the microscopic level, each of these forces will be derived from a potential energy function. Our final expression for the total energy, including the internal energy of the target body, is

$$
\begin{aligned}
E_{\text{total}} \;=\; & \tfrac{1}{2}m\dot{x}^2 + \tfrac{1}{2}(M/2)(\dot{y}^2 + \dot{z}^2) \\
& + \tfrac{1}{2}(M/4)\omega_0^2(y-z)^2 + \tfrac{1}{2}M\omega_0^2\ell^2[\ell/(x-y)]^n/n \quad .
\end{aligned}
\tag{8.23}
$$

The first three terms here are simply the kinetic energies of the three point particles that make up our system; the last two terms are potential energies corresponding to the two forces that we will now discuss.

Model Forces First consider the internal force that binds y to z. The potential energy we chose for it was $U_{\text{int}} = \tfrac{1}{2}\mu_{\text{int}}\omega_0^2(y-z)^2$. The constants need a lot of explaining (or a haughty disdain since they are not the main point). The principal feature of this potential energy is that it gets large when y and z are far apart, so that it requires a very large energy to pull the target apart into its separate particles. It is in this sense that the target is a single object. (Any positive even power of $(y-z)$ has this property, a quadratic is the lowest power we can choose.) The force on the particle at y then is just $-dU_{\text{int}}/dy = -\mu_{\text{int}}\omega_0^2(y-z)$ and is a force of the linear type we used in the **LEAP** worksheet in Section 4.2 on page 65. (The constant $\mu_{\text{int}} = M/4$ is the reduced mass (see Section 8.4) of the two interacting $M/2$ masses; ω_0 controls the strength of this force and has the dimensions of a frequency or reciprocal time. In the worksheet computations we choose units where M is the unit of mass and $1/\omega_0$ is the unit of time. Then this potential energy for the internal force is calculated as just $(y-z)^2/8$.)

For a force that makes the target repel the projectile, we want its potential energy to be large whenever the projectile at x gets close to the nearer target particle at y. This is done by using a negative power $-n$ of $(x - y)$. For convenience, we introduce a constant ℓ to specify a distance beyond which this second force becomes insignificant. The potential energy term $U_{\text{ext}} = \frac{1}{2}M\omega_0^2\ell^2[\ell/(x - y)]^n/n$ in equation 8.23 leads to a force on the particle at y of $-dU_{\text{ext}}/dy = -M\omega_0^2\ell[\ell/(x - y)]^{n+1}$ and an equal and opposite force on the projectile at x.

Given these forces, represented schematically in Figure 8.4, the Newton

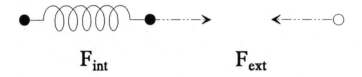

$$\mathbf{F}_{\text{int}} \qquad\qquad \mathbf{F}_{\text{ext}}$$

Figure 8.4: Inelastic collision model, forces.

equations of motion for the three particles solved by the worksheet INELC are

$$m\ddot{x} = +\frac{M}{2}\omega_0^2\ell\left(\frac{\ell}{x - y}\right)^{n+1} \tag{8.24}$$

$$\frac{M}{2}\ddot{y} = -\frac{M}{2}\omega_0^2\ell\left(\frac{\ell}{x - y}\right)^{n+1} - \frac{M}{4}\omega_0^2(y - z) \tag{8.25}$$

$$\frac{M}{2}\ddot{z} = +\frac{M}{4}\omega_0^2(y - z) \tag{8.26}$$

but in the worksheet the choice of units lets us set $M = 1$ (i.e., M is the unit of mass) and $\omega_0 = 1$ (i.e., $1/\omega_0$ is the unit of time). Note that by adding these three equations one finds $m\ddot{x} + \frac{M}{2}(\ddot{y} + \ddot{z}) = 0$ or $m\ddot{x} + M\ddot{w} = 0$ which is a statement of conservation of total momentum $P_{\text{tot}} = m\dot{x} + M\dot{w}$ for this system.

Explore

The worksheet in the student software package is set up to make the target always at rest initially with $w = 0 = \dot{w}$; it tries to choose the initial position x of the projectile and the time step dt, so that the collision with the target will occur near the midpoint of the time interval calculated and presented on the graphs. In addition to the construction parameters m/M, ℓ, and n that appear in Newton's equations for this model, you may also choose as initial conditions the initial values of an energy E_{colln}, which is the kinetic energy in the relative motion $(x - w)$ of the two bodies, and initial values of an internal energy E_{int}, which is associated with the relative motion $(y - z)$ internal to the target body. An additional descriptor ϕ adjusts the initial ratio of kinetic and potential energy in the internal energy; it is typical of microscopic variables that are almost never measured in experiments. (When in large systems there are many such variables, experimentally significant predictions can be made based on their statistical behavior.) The mathematical definitions of these quantities can wait until you look at a number of solutions to see the possible ranges of behavior.

As you explore possible behaviors, any one of the three quantities E_{colln}, E_{int}, or ℓ can fix the unit of length and need not be varied. A graph plotting x, w, and y shows the motions taking place in the collision; another graph plots the projectile kinetic energy $E_x = \frac{1}{2}m\dot{x}^2$, the kinetic energy $E_w = \frac{1}{2}M\dot{w}^2$ in the target's center of mass motion, and E_{int} to allow you to follow the energy exchanges during the collision. Find examples of inelastic collisions where the internal energy is different before and after the collision. More difficult perhaps, but quite possible, is to find conditions under which the collisions are reliably (independent of ϕ) elastic to good accuracy. (Note: the worksheet uses a time stamp to write an identification `Data set #`... on the I/O screen and on each graph. This ID number changes when you press {Calc}, so use {Calc} only before you begin to print a related set of I/O data screen and graphs.) When you have seen many collision examples and followed the energy transfers that occur, you will be ready to see how these different energies are extracted from the basic formula $E_{\text{total}} = \text{KE} + \text{PE}$ of equation 8.23.

Varieties of Energy

When this model is viewed as two colliding bodies, the important energies are the projectile's kinetic energy $E_x = \frac{1}{2}m\dot{x}^2$ and the kinetic energy in the target's center of mass motion $E_w = \frac{1}{2}M\dot{w}^2$. To extract these from the full energy expression in equation 8.23, the coordinate $w = (y+z)/2$ of the center of mass of the target body is introduced by writing $2(\dot{y}^2 + \dot{z}^2) = (\dot{y} + \dot{z})^2 + (\dot{y} - \dot{z})^2 = 4\dot{w}^2 + (\dot{y} - \dot{z})^2$. Then equation 8.23 can be rewritten as

$$
\begin{aligned}
E_{\text{total}} \;=\; & \tfrac{1}{2}m\dot{x}^2 + \tfrac{1}{2}M\dot{w}^2 \\
& + \tfrac{1}{2}(M/4)[(\dot{y} - \dot{z})^2 + \omega_0^2(y - z)^2] \\
& + \tfrac{1}{2}M\omega_0^2\ell^2[\ell/(x - y)]^n/n
\end{aligned}
\tag{8.27}
$$

in which we identify

$$
E_{\text{int}} = +\tfrac{1}{2}(M/4)[(\dot{y} - \dot{z})^2 + \omega_0^2(y - z)^2]
\tag{8.28}
$$

as the internal energy terms. If $\ell = 0$ so there is no interaction with x, this energy will be conserved by itself, as will $E_x = \frac{1}{2}m\dot{x}^2$ and $E_w = \frac{1}{2}M\dot{w}^2$. But that means that it will also be conserved when the two bodies x and w are far apart, i.e., before and after the collision.

Chapter 9

Energy Dynamics

There are three points covered in this chapter. First, if you have studied rigid body motion in your text, this chapter provides problems that apply those principles to the analysis of some mechanical motions and to their modeling with spreadsheets. (If your course omits rigid body dynamics to allow time for computational physics then similar problems can be formulated using point particles instead.) The second point covered in this chapter is that, for motions with a single dynamical coordinate, the content of Newton's laws can be recovered from the energy equation. This technique avoids many difficulties inherent in a more direct application of Newton's laws. The third point illustrated here is that you can use conventional units on the I/O page, while letting the computations in the work block be simplified by using dimensionless variables there. In the **ROLL** worksheet developed in this chapter, using conventional units for the problem description on the I/O page allows certain quantum mechanical limitations to be checked. These depend on the scale of the apparatus, although Newton's laws lead to similar motions on all scales as reflected in the solution we construct using dimensionless quantities.

9.1   Newton's Laws from Energy Conservation

When energy is conserved—and when there is only one independent variable needed to describe a motion—a common and efficient way to find an equation for the acceleration of that variable is to differentiate the energy equation. The simplifications achieved are often substantial. The exercises at the end of this section provide many examples. The simplest illustration of this technique is the case of a particle moving in one dimension under the action of a conservative force with potential energy $U(x)$. The conserved total energy then is

$$E = \tfrac{1}{2}m(dx/dt)^2 + U(x) \qquad (9.1)$$

and, since E is constant, the derivative of this equation is

$$0 = (dx/dt)[m(d^2x/dt^2) + dU/dx] \quad . \qquad (9.2)$$

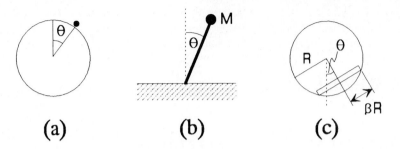

Figure 9.1: The energy dynamics approach can be applied to the sample
mechanisms sketched here. Figure (a) is a point particle sliding without
friction on a cylinder, a simplified version of Figure 9.3, which has a ball
instead of a point mass. Figure (b) represents a point mass at the end
of a rigid, massless stick, which may slide on the horizontal surface if
friction is insufficient. In figure (c) we have a hollow cylinder and, inside
it, a rigid bar placed crosswise to the cylinder axis and able to slide
without friction in the plane of the diagram.

There are two mathematical solutions to this equation: either the particle
doesn't move, $dx/dt = 0$, which usually violates Newton's laws (but doesn't
violate conservation of energy), or $[\ldots] = 0$, which is just Newton's law in this
case,

$$m(d^2x/dt^2) = -dU/dx \quad .$$
(9.3)

In applying this approach, you must always be careful to ignore the unphysical
(no motion) solution and use the solution that yields an acceleration equation.

A more typical application of this energy method is illustrated in the ROLL
worksheet described in the next section. A simplified version of it considers a
small mass M that slides without friction on a cylinder whose axis is horizontal,
as suggested in Figure 9.1(a). The questions that arise include: How long can
the small mass balance on the cylinder before it falls off? Through what angle
will the mass slide around the cylinder (from near the top) before it flies off?
These questions are answered from a spreadsheet model that starts from the
energy method. Since we have assumed no friction, energy will be conserved.
To calculate the energy we assume a radius R for the cylinder and a second
coordinate θ that measures the angular position around the surface of the
cylinder, measured down from the top. For simplicity we will take the velocity
of the mass along the axis of the cylinder, v_z, to be zero. The motion in the xy-
plane (perpendicular to the horizontal z-axis) will be circular with a velocity
$R\dot\theta$, so the kinetic energy will be $\frac{1}{2}v^2 = \frac{1}{2}(\dot x^2 + \dot y^2) = \frac{1}{2}R^2\dot\theta^2$. We must also
include gravitational potential energy $Mgy = MgR\cos\theta$ which is conveniently
measured from the level of the cylinder axis. Thus we have found an energy
formula

$$E = \tfrac{1}{2}MR^2\dot\theta^2 + MgR\cos\theta$$
(9.4)

that only contains one variable, θ. The advantage this approach has over directly formulating Newton's laws is that we did not need to deal explicitly with the normal force, so our equation does not contain this unknown quantity. By differentiating equation 9.4 and discarding the unphysical solution $\dot{\theta} = 0$ we find an acceleration equation

$$0 = MR^2\ddot{\theta} - MgR\sin\theta \qquad (9.5)$$

or

$$\ddot{\theta} = +(g/R)\sin\theta \quad . \qquad (9.6)$$

This equation can be readily solved with a spreadsheet and allows us to graph $\theta(t)$. It does not yet answer the questions we posed, however, since they require knowledge of the normal force.

Equations 9.4 and 9.6 assume that the mass is sliding on the surface of the cylinder, but do not tell what normal force is required to keep it there. We may assume that the cylinder is rigid enough to supply any required push (normal force N positive) to keep the mass from denting the surface, but we do not want to assume that it can pull the mass down to the surface when the momentum of the mass wants to carry it away. It is therefore necessary to write an equation using Newton's second law to find the normal force. But with this energy method, we get a solvable equation in one unknown first (equation 9.6). We use Newton's law only later to find unknown forces when that is required. To this end, we write the vector component of Newton's second law in the direction normal to the surface. The acceleration in this direction is the centripetal acceleration in the circular motion $a_n = -R\dot{\theta}^2$, and the force is the normal force N of the cylinder pushing against the moving mass plus the component of gravity in this (normal) direction $-Mg\cos\theta$. Thus we have

$$Ma_n \equiv -MR\dot{\theta}^2 = N - Mg\cos\theta \qquad (9.7)$$

or

$$N/Mg = \cos\theta - (R/g)\dot{\theta}^2 \quad . \qquad (9.8)$$

A worksheet that calculates $\theta(t)$ can also calculate and plot $N(t)$ and thus point out the unphysical portion of the graphs where $N < 0$ signals that the particle will have flown off its cylindrical support.

EXERCISE 9.1^C Draw a free body diagram for the point mass sliding on a cylinder described and shown in Figure 9.1(a) and deduce equation 9.6 directly from Newton's second law.

EXERCISE 9.2^C Idealize a lollypop as a point mass M on the end of a rigid mass-less rod of length L, as sketched in Figure 9.1(b). Try to balance the lollypop on a table top, with the mass at the top. Describe its position by an angle θ measured from the vertical. By using the energy method find an acceleration equation that can be solved on a spreadsheet for $\theta(t)$ while the foot of the lollypop does not slip.

EXERCISE 9.3^P From a free body diagram for the falling lollypop in the previous exercise, find ways to compute the normal force N and the friction force f that act

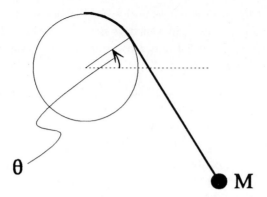

Figure 9.2: Modified pendulum. A flexible cord wraps around a cylinder as the pendulum swings. See Exercise 9.7.

on the foot of the lollypop to keep it from slipping. Give equations that allow these to be computed once $\theta(t)$ is known. This then also gives the minimum coefficient of static friction $\mu_{min} = f/N$ required to prevent slipping at any point in the fall.

EXERCISE 9.4^P Repeat the previous exercise, but now consider a pencil balanced on its point (idealized as a uniform rod of length L) instead of a lollypop.

EXERCISE 9.5^P Inside a fixed hoop of inner radius R (axis horizontal) lies a rigid rod of length $2\beta R$ forming a movable chord of the circle, as sketched in Figure 9.1(c). Find the equation of motion (acceleration equation) for this rod as it slides frictionlessly inside the hoop, with all motion confined to the (vertical) plane of the hoop.

EXERCISE 9.6^P A bead slides frictionlessly on a wire that is formed into a helix described by $z = k\theta$ and $r =$ const where r and θ are polar coordinates in the horizontal xy plane, and z is the vertical direction defined by gravity. Find an equation of motion for $z(t)$ and its analytic solution.

EXERCISE 9.7^H A modified pendulum is constructed as follows: A point mass M is fixed at the end of a perfectly flexible and inextensible cord. The other end of the cord is attached to the top of a fixed horizontal cylindrical bar of radius R. The length of the cord is $R + L$. Find the equation of motion for the angle θ made by the straight lower end of the cord and the vertical, as in Figure 9.2. Identify a set of units (or of dimensionless variables) that simplify this equation.

9.2   ROLL: Ball Rolling off a Cylinder

We now want to complete a worksheet to solve for a motion like that analysed in the preceding section. If you have studied rigid body motion you should be able to calculate the energy when a ball rolls without slipping (instead of a point mass sliding frictionlessly) on the cylinder to find

$$E = \tfrac{1}{2}MR^2\dot{\theta}^2 + \tfrac{1}{2}I\dot{\phi}^2 + Mg(R\cos\theta - R) \qquad (9.9)$$

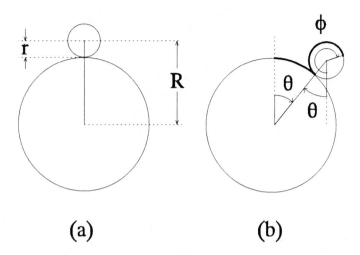

Figure 9.3: Ball rolling on a cylinder. Diagram (a) shows the $\theta = 0$ position; we will follow the vertical mark on the ball to see the angle ϕ through which it turns as it rolls. Diagram (b) shows another position θ after the ball has rolled without slipping. The two emphasized arcs that have contacted each other during the roll must have the same length: $(R - r)\theta = r(\phi - \theta)$.

since $R\dot{\theta}$ is the velocity of the center of mass of the ball. See Figure 9.3. (We assume no motion in the z direction and have set the zero of potential energy at $\theta = 0$.) In this energy expression $\dot{\phi}$ is the angular rotation rate of the ball about its center of mass. The "no slipping" condition relates it to $\dot{\theta}$. From Figure 9.3 you can see that $(R - r)\theta = r(\phi - \theta)$. (Imagine that the ball and cylinder leave marks on each other when they touch.) This simplifies to $R\theta = r\phi$ which implies $r\dot{\phi} = R\dot{\theta}$. Let us now write I in a way that distinguishes the size and the shape of the ball, so we can easily replace it with a hoop or a disk or a hollow ball if we want. This is done by writing $I = \kappa^2 M r^2$, which factors out a typical magnitude $M r^2$ for the moment of inertia and leaves a dimensionless shape-dependent factor κ^2 adjustable. For a sphere one has $\kappa^2 = 2/5$; for a thin spherical shell one has $\kappa^2 = 2/3$. To treat only the point mass case, set $I = 0$ and $\kappa = 0$ in what follows. By using this form of I with $r\dot{\phi} = R\dot{\theta}$ we can rewrite the energy equation as

$$E = \tfrac{1}{2}(1 + \kappa^2)M R^2 \dot{\theta}^2 + Mg(R\cos\theta - R) \quad . \tag{9.10}$$

Note that r has disappeared from the problem. A small ball must turn faster than a large one to roll without slipping at the same speed; we have found that this gives them the same rotational kinetic energy for equal mass. Note also that we have not looked at—and do not need to look at—the friction force that keeps the ball from slipping as it rolls, even though this force significantly influences the tangential acceleration dv_θ/dt of the ball. Its effects are properly included when we derive the acceleration equation 9.15. We do assume, however, that friction is sufficient to prevent slipping.

Dimensional Analysis

Dimensional analysis helps to simplify equations to be solved on a spreadsheet by expressing the most important variables of the problem in dimensionless form. It shortens the computations, so that the worksheet recalculates more quickly when data are changed. It also allows different quantities (position, velocity, energy, etc.) to be plotted together on a single graph, since when made dimensionless these are more likely to have comparable magnitudes. Previously we realized the advantages of dimensionless variables by choosing units flexibly. Here we could obtain simplifications by letting M be the unit of mass (set $M = 1$) and R be the unit of length (set $R = 1$). But this time we want to retain conventional SI units on the I/O page so that some quantum mechanical criteria can be discussed later. Therefore we use a different approach here. We write equation 9.10 in dimensionless form by dividing by an energy constant that is a typical amount of energy for this problem; we choose this unit to be MgR and write

$$E/MgR = \tfrac{1}{2}(1 + \kappa^2)(R/g)\dot{\theta}^2 + (\cos\theta - 1) \quad . \tag{9.11}$$

The potential energy term is clearly dimensionless on the right hand side of this equation, but the kinetic energy term can be improved. Since θ is dimensionless, the coefficient must cancel the time units in the denominator of the derivatives. It is useful to think of this coefficient as defining a natural frequency for the problem:

$$\omega_0^2 = \frac{g/R}{1 + \kappa^2} \tag{9.12}$$

and a corresponding dimensionless time coordinate

$$\tau = \omega_0 t \quad . \tag{9.13}$$

With these definitions, equation 9.11 can be rewritten as

$$E/MgR = \tfrac{1}{2}(d\theta/d\tau)^2 + (\cos\theta - 1) \quad . \tag{9.14}$$

We will use an energy equation in this form in the work block of our worksheet. The column headings in Screen 9.1 show this by, for instance, indicating in the **energy** column that the units used are MgR, implying that the numbers in that column are the values of E/MgR.

   An equation of motion for θ can be found by differentiating equation 9.14 with respect to τ. The result, ignoring a common factor $d\theta/d\tau$, is

$$0 = \frac{d^2\theta}{d\tau^2} - \sin\theta \quad . \tag{9.15}$$

	J	K	L	M	N	O	P	Q
			J	**K**	**L**	**M**	**N**	**O**	**P**	**Q**

	J	K	L	M	N	O	P	Q
97	name:	time	angle	ang_vel	ang_accel	energy	ang_vel	norm_fce
98	units:	[1/wo]	[1]	[wo]	[wo^2]	[MgR]	[wo]	[Mg]
99	labels:	t	q	whs	a	E	w	N
100	init data:	0.000	0.010	0.000	0.010	-0.000	0.000	1.000
101	typical line:	0.026	0.010	0.000	0.010	-0.000	0.000	1.000
102	copied lines:	0.052	0.010	0.001	0.010	-0.000	0.001	1.000
103		0.079	0.010	0.001	0.010	-0.000	0.001	1.000
104		0.105	0.010	0.001	0.010	-0.000	0.001	1.000
105		0.131	0.010	0.001	0.010	-0.000	0.001	1.000
106		0.157	0.010	0.002	0.010	-0.000	0.002	1.000

Screen 9.1: Beginnings of the work block of ROLL.wk1.

In Exercise 9.8, you will derive the equation

$$N/Mg = \cos\theta - (R/g)\dot{\theta}^2 \quad . \tag{9.16}$$

for the normal force. When expressed in terms of our dimensionless time τ this becomes

$$\frac{N}{Mg} = \cos\theta - \frac{1}{1+\kappa^2}\left(\frac{d\theta}{d\tau}\right)^2 \tag{9.17}$$

and gives a dimensionless measure N/Mg of the normal force.

Completing the Worksheet

In the student software package is a nearly complete worksheet ROLL.wk1 with the I/O page and work block column headings already laid out. The start of its work block and its home screen are shown in Screens 9.1 and 9.2. (It is based on the **LEAP** worksheet from Section 4.2 since the equation to be solved, equation 9.15, has the same general form as the equation $\ddot{x} = F(x)/m$ solved there.) You need only complete the work block, using the dimensionless equations we have just developed.

The I/O page contains some comparisons to quantum mechanical limits discussed in optional Section 9.4. It also simplifies the choice of the numerical time step **dt** by letting you set it as some fraction of a time t_{crit} that is an analytic estimate of the time required for the ball to reach a velocity where it flies off the cylinder. The basis for this estimate is given in optional Section 9.3. As a further convenience, the notes page of the **ROLL** worksheet reports the **Fall time** and **Fall angle**, which are the time and θ value at which the normal force drops to zero. You can, of course, find these values yourself by scanning the work block tabulation to find the row in which the normal force first goes negative and noting the time and angle there. In a real mechanism our equations fail to apply after this time, and the correct equations thereafter

F8: PR [W10] "   h = Planck's constant   READY

	A	B	C	D	E	F	G	H
1	your name & today's date							
2	your class & section			ROLL, unstable equilibrium				
3				Uses the leapfrog integration method				
4	--							
5	Construction parameters:							
6	k =	0.632	r	= gyration radius;	1+(k/r)^2=		1.400	
7	g =	9.8	m/s^2	= Earth gravity;		M =	0.050	kg
8	R =	0.30	m	= total radius;		h =	6.63E-34	J s
9	Initial conditions:							
10		q_init =	0.01	radian	initial angle away from equilibrium			
11		v_init =	0	m/s	initial velocity			
12	Quantum limit:		3.4E+28	> 1 ?	Can't have q_init, v_init too small			
13	Approximation:							
14		dt =	0.006	s	=	0.005	*t_crit =	0.026 /wo
15	--							
16	Notes page				Fall angle:		0.949 rad	
17	Freq. unit, wo		4.082	rad/s	Fall time:		1.290 s	
18	Time unit, 1/wo		0.245	s			criterion	
19	Crit time, t_crit		1.283	s			t	
20	Crit ang, q_crit		0.942	rad			+N<0	

ROLL.WK1

Screen 9.2: Home screen of the ROLL worksheet.

are be those of free fall—without a negative normal force the ball cannot be kept in contact with the cylinder but continues on a parabolic trajectory.

Exploration

Try various initial conditions for the ROLL worksheet after you have completed it. You should find (in Exercise 9.11) that the fall time depends only weakly on the initial angle θ_0 when the ball starts at rest near the top, and that the fall angle (Exercise 9.12) is essentially independent of θ_0 for small θ_0. What is the largest initial velocity $v_0/\omega_0 R$ (dimensionless) that you can specify at $\theta_0 = 0$ without needing a negative N at time zero to keep the ball on the cylinder? Try starting at $\theta_0 < 0$ with a positive v_0 to roll the ball over the top. You should find that the fall angle is then always greater than $|\theta_0|$. Can you prove analytically (see Exercise 9.14) that this must be so?

EXERCISE 9.8[C]   Draw a free body diagram for a ball rolling on a cylinder, as shown in Figure 9.3, and explain how Newton's second law leads to equation 9.16.

EXERCISE 9.9[C]   Let a ball of radius r roll without slipping on the inside surface of a horizontal hollow cylinder of radius $R + r$. Derive, by the energy method, a differential equation analogous to equation 9.6 for the angular position θ of the ball's center.

EXERCISE 9.10[C]   Complete the ROLL worksheet and print out the formulae from the first two rows of computation in the work block. (If you wish to save the completed worksheet in compacted form, see Appendix C, Section C.1, for the use of the {Alt-S} macros that are provided in the distributed ROLL worksheet.)

EXERCISE 9.11^{E} On the notes page of your worksheet, make a table of fall times and fall angles for a variety of initial angles θ_0 with $v_0 = 0$. (Run the worksheet many times and copy the reported results to your table using the /Range Value command.) Include some very small initial angles. Graph these fall times and fall angles as functions of θ_0. Also make semilog plots of them by adding columns to the table that tabulate $\log_{10}(t_{fall})$ and $\log_{10}(\theta_{fall})$. Print these graphs.

EXERCISE 9.12^{E} Calculate analytically, using energy arguments, the fall angle for initial conditions with given dimensionless energy $\epsilon = E/MgR$. How does it depend on ϵ for very small ϵ?

EXERCISE 9.13^{E} Calculate analytically the largest initial velocity $v_0/\omega_0 R$ at $\theta_0 = 0$ that is consistent with a positive normal force N. (Check your work by inspecting the worksheet computations for values on both sides of your analytic result. If the numerical and analytical approaches do not agree, find out which one is wrong and correct it.)

EXERCISE 9.14^{E} When θ_0 is negative and v_0 is positive the ball rolls up and over the top of the cylinder before rolling down the opposite side. Naturally, v_0 must be large enough to allow the ball to reach the top yet small enough so that the ball remains in contact with the cylinder. Under these circumstances prove analytically that $\theta_{fall} \geq |\theta_0|$.

EXERCISE 9.15^{P} Modify the ROLL worksheet to treat the lollypop problem (Exercises 9.2 and 9.3). Compare the $\ddot{\theta}$ equations in the two cases (lollypop and ball rolling on a cylinder) to see how both the choice of natural units (e.g., ω_0) and the dimensionless equations may need changes. Insert a column that calculates the coefficient of static friction required to prevent slipping at each time step and let this replace the normal force on the main graph. (Note: if you use the {Alt-S} macro in the distributed ROLL worksheet to save your modifications, see Appendix C, Section C.1. New columns may be inserted only within the work block, not added at its edges; the cell SAVENAME must contain your new filename, e.g. LOLLY.)

EXERCISE 9.16^{P} Repeat the previous exercise, but now consider a pencil balanced on its point (idealized as a uniform rod of length L) instead of a lollypop.

EXERCISE 9.17^{P} Modify the ROLL worksheet to treat the sliding rod problem (Exercise 9.5). Compare the $\ddot{\theta}$ equations in the two cases (sliding rod versus ball rolling on a cylinder) to see how the choice of both natural units (e.g., ω_0) and the dimensionless equations may need changes.

EXERCISE 9.18^{H} Modify the LEAP or ROLL worksheet to solve the motion analyzed in Exercise 9.7. Check the worksheet by comparing the small angle motion with an analytic solution using the approximations $\sin \theta = \theta$ and $\cos \theta = 1$. Print phase-plane plots of the oscillation at several different amplitudes.

9.3 ♠ Analytic Approximations

In a small angle approximation, equation 9.15 can easily be solved. This lets us estimate the time required to reach a given angle such as the fall angle. The approximation we use when θ is small is

$$\sin \theta \approx \theta \qquad (9.18)$$

so that equation 9.15 becomes

$$\frac{d^2\theta}{d\tau^2} = \theta \qquad (9.19)$$

whose solutions are

$$\theta \propto \exp(\pm\tau) = \exp(\pm\omega_0 t) \qquad (9.20)$$

or some linear combination of these to fit the initial conditions. In fact

$$2\theta = (\theta_0 + \theta_0')e^{\tau} + (\theta_0 - \theta_0')e^{-\tau} \qquad (9.21)$$

has the specified initial conditions where $\theta_0' \equiv v_0/\omega_0 R$. For $\tau \equiv \omega_0 t \gg 1$ this equation can be well approximated by

$$\theta = \tfrac{1}{2}(\theta_0 + \theta_0') \exp(\omega_0 t) \qquad (9.22)$$

which can be solved for t:

$$t = \frac{1}{\omega_0} \ln \left[\frac{2\theta}{\theta_0 + \theta_0'} \right] \qquad . \qquad (9.23)$$

This is used in the worksheet as an estimate for t_{fall} by supplying the value of θ_{fall} from Exercise 9.12. It should be exact only when $|\theta_{\text{fall}}| \ll 1$ and $\omega_0 t_{\text{fall}} \gg 1$.

9.4 ♠ Quantum Considerations

The unstable phenomena treated in this chapter provide a tantalizing challenge to test quantum mechanics. When a pencil is balanced on its tip or a ball on the top of a hill, very small disturbances will make it fall. Quantum mechanics contains an uncertainty principle that puts limits on how small such disturbances can be. You disprove quantum mechanics if you succeed[1] in balancing a pencil on its tip for more than several seconds!

---

[1] Of course any such experiment would be subject to close scrutiny: was there a dent on the top of the experimental cylinder, inconsistent with the theoretical model, that would stabilize the ball? or a magnetic impurity that held it in place? was there a flat spot at the tip of the pencil that was not included in the theoretical calculations? etc.

The Heisenberg Uncertainty Principle

Fundamental to quantum mechanics are limitations on the meaningfulness of the usual (Newtonian) mechanical concepts of "initial conditions" or the "state of the system" that we have been using. These limitations make an atom somewhat like a hazy malleable ball and quite unlike the miniature solar system that the older Bohr model of the atom envisioned.

Initial conditions are sets of data such as the (θ_0, v_0) used in several examples in this chapter or the (x_0, v_0) specified to begin integration of many one-dimensional motions. As we have seen in phase-plane plots, such a pair (x, p) defines a state from which the subsequent states are completely determined by the Newtonian equations of motion. Quantum mechanics is written in a mathematical language of wave equations and matrices that makes it impossible to assign definite values to such data sets. Heisenberg's uncertainty principle expresses this limitation in a simple way: $\Delta x \Delta p \geq \frac{1}{2}\hbar$ where $h \equiv 2\pi\hbar = 6.6 \times 10^{-34}$ Joule-seconds is known as Planck's constant. ($\hbar$ is simply called "h-bar.") In this statement, the Δx and Δp are the uncertainties in the specifications of position and momentum; if either x or p is known well, the other must be more uncertain to keep the uncertainty product above the Heisenberg limit. Many pairs of such quantities, known as complementary or conjugate pairs, satisfy relations of this form. For the examples in this chapter the relationship is $\Delta\theta\Delta J \geq \frac{1}{2}\hbar$ where J is the angular momentum.

The uncertainty condition $\Delta x \Delta p \geq \frac{1}{2}\hbar$ can be interpreted in the phase plane as saying that points there are unrealizable—the best one can do is to identify an area $\Delta x \Delta p$ not smaller than $\frac{1}{2}\hbar$ in size to characterize the state of the system.

Quantum Uncertainties in ROLL

In this chapter's main example, the ball rolling off a cylinder, your worksheet should predict rather large fall times if you set $v_0 = 0$ with θ_0 sufficiently small. (How small must θ_0 be to produce a fall time of 30 seconds?) The uncertainty principle will show that such long times are unattainable in principle and not just because of deficiencies in your experimental technique in placing the ball at the exact top of the cylinder. For a long fall time the desired initial condition would be to have both θ_0 and v_0 be as close to zero as possible. We can regard any nonzero θ_0 or v_0 we encounter as a necessary uncertainty and look for motions where the initial values of θ_0 and J_0 are restricted by the Heisenberg principle. We therefore want to check our initial conditions against the requirement that $|\theta_0 J_0| \geq \frac{1}{2}\hbar$.

The angular momentum J in our example, measured about the center of the cylinder from which θ is measured, is given by

$$\begin{aligned} J &= MR^2\dot{\theta} + I\dot{\phi} \\ &= MR^2(1 + \kappa^2)\dot{\theta} \quad . \end{aligned}$$

(9.24)

$$(9.25)$$

For the simpler problems of a point mass sliding on the cylinder, or the idealized lollypop in Exercise 9.2, you can set $I = 0 = \kappa^2$. The first term here arises from the motion of the center of mass of the ball; the second term from the ball's spin about its center of mass. The Heisenberg uncertainty limits our initial values for θJ, which has units of energy $\times$ time (Joule-seconds), which is called action. Multiplying the condition by ω_0 restates it in more familiar energy units: $\omega_0 \theta_0 J_0 \geq \frac{1}{2}\hbar\omega_0$. A bit of fancy algebra, using estimates like $|2ab| \leq a^2 + b^2$, which follows from $(a \pm b)^2 \geq 0$ and $\cos\theta \approx 1 - \frac{1}{2}\theta^2$ for small θ, allows one to show (Exercise 9.20) that if

$$E \geq \tfrac{1}{2}\hbar\omega_0 \tag{9.26}$$

is violated, then so is the Heisenberg relationship

$$|\theta J| \geq \tfrac{1}{2}\hbar \quad . \tag{9.27}$$

It is inequality 9.26 that we check on the I/O page of the ROLL worksheet. For values of M, R, and κ^2 that correspond to some apparatus you can experiment with, find the longest fall time quantum mechanics allows by trying some very small values of θ_0 on your spreadsheet.

Although it is not as general as the uncertainty condition $\Delta x \Delta p \geq \frac{1}{2}\hbar$, the energy condition 9.26 is valid for many simple systems and shows that microscopic motion does not cease even at absolute zero temperature. The uncertainty principle is fundamental to the stability of atoms and other matter; without it atoms would collapse in a flash and the human spirit would have neither flesh nor paper with which to write its poetry.

EXERCISE 9.19^{C} Specify practical values for the construction parameters in the ROLL worksheet and find the corresponding longest fall time allowed by the quantum uncertainty principle. What is the longest fall time you have been able to achieve experimentally (e.g., with a pencil at your desk) for roughly comparable parameters?

EXERCISE 9.20^{H} Prove the relationship between inequality 9.26 and the Heisenberg uncertainty principle 9.27. Start by showing that

$$\omega_0 \theta J = MgR\theta\theta' \tag{9.28}$$

where $\theta' \equiv d\theta/d\tau$, using the definitions given for J and ω_0. Then use a small angle approximation for $\cos\theta$ in equation 9.14 to simplify the energy formula and apply the $|ab|$ inequality 9.27.

Chapter 10

Gravity and Central Force Orbits

This chapter provides a series of three worksheets for examining motion of an object under the influence of a central force. The first of these, CENTF.wk1, models the motion of a mass moving about a center of force whose attraction varies as an inverse power of the object's distance from the center. For the inverse-square-law case, the force behaves like the gravitational attraction between the sun and a single planet or comet. In this case the worksheet should reproduce Kepler's three laws of planetary motion. CENTF.wk1 also lets you explore what happens when forces differ from inverse-square-law behavior. The second worksheet, RSCAT.wk1, adapts CENTF.wk1 to the study of the deflection of rapidly moving objects as they pass close to such a force center. A nonperiodic comet passing near the sun or a negatively charged elementary particle passing close to an atomic nucleus are two examples of this situation. The main goal here is to discover how close the object must pass to the force center in order to be deflected significantly. The third worksheet, SLING.wk1, invites you to try your hand at navigating a spacecraft on a journey outward from the sun so that the spacecraft gets a gravitational boost from a massive outer planet. A close encounter with the planet, properly timed, will assure the success of your mission.

10.1   CENTF: Kepler Motion

The gravitational force attracting one mass m toward another mass M is $F(r) = GMm/r^2$ directed toward M. If M is fixed at the origin of the coordinate system used to describe the motion of m, then this force always points radially inward. Take a moment now to convince yourself that the trajectory of m as it moves about M lies in the plane determined by $\mathbf{F}$ and $\mathbf{v}$ (evaluated at any time).

141

From a careful analysis of the observed motion of the planets, Johannes Kepler (1571–1630) concluded that their motion satisfied three empirical laws.

1. All planets move in elliptical orbits with the sun at one focus of the ellipse.

2. The radius vector from the sun to the planet sweeps out equal areas in equal times.

3. The square of the period of the planet's orbit is proportional to the cube of the orbit's semimajor axis.

Newton showed that planets attracted towards the sun by a force $F(r) = GMm/r^2$ obey Kepler's three laws. You should check that the orbits that your completed worksheet calculates are consistent with Kepler's laws.

While radar measurements of the motion of the planets around the sun have confirmed the inverse-square-law character of the gravitational force to a high precision, it is interesting to speculate how a body might move about the sun if the gravitational force varied as $F(r) = GMm/r^n$ for $n \neq 2$. By making n an adjustable parameter in your worksheet you can easily determine the observable effects of deviations from exact inverse-square-law behavior.

To carry out the calculation of the two dimensional motion of m about M conveniently you will need to express $\mathbf{F}$ in terms of its Cartesian components F_x and F_y. With x and y the coordinates of m, $r = \sqrt{x^2 + y^2}$ is its distance from M, $F = GMm/r^n$ is the magnitude of the force acting on m, and the required vector components are:

$$F_x = -F\frac{x}{r}$$
$$F_y = -F\frac{y}{r} \quad .$$

$$(10.1)$$

Choosing Units

As always, we choose our units both to simplify our computational formulae and to produce numerical values that are easily visualized. Since we have in mind the motions of a planet about a star, we may as well choose the earth's year (1 yr) as our unit of time and the mean radius of the earth's orbit as our unit of distance. This distance unit, termed an AU (astronomical unit) equals 499 light seconds. We can also choose the earth's mass as our unit mass. This choice of $m = 1$ will simplify our treatment of the planet's angular momentum and energy. Since the trajectory and the time scale of the planet's motion do not in fact depend on the planet's mass, (m cancels from Newton's law $ma = -GMm/r^2$) this choice of mass unit is merely a convenience to allow writing E and L instead of E/m and L/m; we can say, for example, energy while we compute E/m.

Setting Up the Worksheet

Like DRAG2.wk1, which you developed in Chapter 3, the CENTF worksheet is laid out to solve for and display all the important kinematical variables for a point mass moving in a plane. It differs from DRAG2 in two ways. The force acting on the mass is now an attractive central force, always directed toward the origin of the coordinate system and dependent only on the distance of the mass from the origin. In addition, it uses the more efficient leapfrog method, first developed in Chapter 4, to integrate Newton's second law.

The worksheet CENTF takes as its starting point CBOND.wk1, introduced in Chapter 5 as our first application of the leapfrog integration technique. As provided in the student software package, CENTF.wk1 already has all necessary column headings and graphs and many of the required formulae prepared for you, but it lacks the formulae and the proper initial value references for the y component of the motion. It still has the expression for the one-dimensional force and energy used in CBOND. As you make the necessary changes, follow the advice you were given in Chapter 8 for a similar project (One Mistake at a Time, page 105).

To complete CENTF you must:

- copy the four columns of formulae for the x motion over to the corresponding four columns describing the y motion;

- replace the old CBOND formulae for force with appropriate expressions for the x and y components of the central force;

- edit the cells in the init data row of the work block so that they refer to the appropriate cells in the home block;

- supply formulae in the remaining four blank columns for the planet's angular momentum L, radial momentum p_r, speed v, and azimuth ϕ; and

- correct the formula for the total energy of the planet by replacing the old potential energy term used in CBOND with the potential energy consistent with the force law you are using and by generalizing the expression for the kinetic energy from one to two dimensions.

In standard algebraic notation, the required formulae are:

$$
\begin{aligned}
a_x &= -x(F/r) \\
a_y &= -y(F/r) \\
(v_{x,y})_n &= \tfrac{1}{2}[(v_{\text{xhs,yhs}})_n + (v_{\text{xhs,yhs}})_{n-1}] \\
&= \tfrac{1}{2}[(v_{x,y})_{n+1/2} + (v_{x,y})_{n-1/2}] \\
r &= \sqrt{x^2 + y^2} \\
(F/r) &= GM/r^{(n+1)} \\
E &= T + U
\end{aligned}
\qquad (10.2)
$$

$$
\begin{aligned}
\text{where} \qquad T &= \tfrac{1}{2}[v_{\text{xhs}}(v_{\text{xhs}} - a_x\,dt) + v_{\text{yhs}}(v_{\text{yhs}} - a_y\,dt)] \\
\text{and} \qquad U &= -[r^2/(n-1)](F/r) \\
L &= xv_y - yv_x \\
p_r &= (xv_x + yv_y)/r \\
v &= \sqrt{v_x^2 + v_y^2} \\
\phi_n &= \phi_{n-1} + dt\,(L/r^2)/2\pi \\
\text{with} \qquad \phi_{\text{init}} &= \arctan(y_{\text{init}}/x_{\text{init}})/2\pi \quad .
\end{aligned}
$$

While you are sorting out the meaning of each of these equations, remember that we have set the planet's mass $m = 1$ so m does not appear explicitly in expressions for the kinetic energy T, the potential energy U, the angular momentum L, the radial component of the linear momentum p_r, or the quantity (F/r). Notice that the geometrical mean of the half-step velocity components is used in calculating the planet's kinetic energy. In Chapter 6 we found (Exercise 6.5) that, in circumstances where the leapfrog method is highly accurate, the geometric mean gives a more constant numerical value for the total energy E than does the arithmetic mean. We therefore expect its constancy to be a sensitive diagnostic for checking the validity of the numerical integration. Also the approximation we choose for angular momentum, $L_n = x_n(v_y)_n - y_n(v_x)_n$, is remarkably sensitive to any failure to program a central force properly.

The reason that (F/r), and not the physically more natural quantity F, is computed is for numerical efficiency. This worksheet takes some time to recalculate, so saving hundreds of multiplications can shorten the time. Similarly, you should only use the formula in equation set 10.2 for the kinetic energy T in the initial row only; thereafter the computation $T_n = [(v_x)_{n+1/2}(v_x)_{n-1/2} + (v_y)_{n+1/2}(v_y)_{n-1/2}]/2$ is more efficient. To save additional computation time, the constants 2π, $(n-1)$, and $(n+1)$ are each computed once in a scratch area and then referenced in formulae in the work block, rather than being recomputed each of the hundreds of times they are used in the work block.

The calculation of the azimuthal position of the planet ϕ, as presented, may be difficult to understand at first. Its purpose in CENTF.wk1 is to note

```
D2: PR [W8] +"Central force, 1/r^"&@STRING(N,1)                    READY
```

```
         A        B        C       D       E      F      G      H
 1   your name  &  today's date
 2   your class & section          Central force, 1/r^2.0
 3                                         Leapfrog integration
 4   --------------------------------------------------------------------
 5   Construction parameters:                          #2pi = 2*@PI
 6          Central Mass  GM =   +#2PI^2                n-1 =      1
 7          Force exponent n =     2.0                  n+1 =      3
 8                    F = -GM/r^n
 9   Initial conditions:            t_init =   0.000
10          x-Position:  1.100      x-Veloc:       0
11          y-Position:  0.000      y-Veloc: 0.903*#
12   Approximation:
13              dt =     0.015
14   Results:
15          r_min =       0.90 r_max =    1.10 r_mid =   1.00
16          r_period [time]  =   1.00        eccen=   0.10
17          r_period [rev]   =   1.00     time/rev=   1.00
18   --------------------------------------------------------------------
19   Notes page:        Use Alt-M for More integration
20                      (continues from previous endpoint).
     CENTF.WK1
```

Screen 10.1: The home screen of the CENTF worksheet.

when the planet has completed each full revolution of its orbit. By dividing the usual angular position of the planet by 2π, we have converted the units of ϕ from radians, as assumed in standard trigonometric formulae, to revolutions. The function $\arctan(y/x)/(2\pi)$ seems to do the desired job, provided you remember to use the four-quadrant form of the spreadsheet function @ATAN2(X_INIT,Y_INIT)/(2*@PI), which correctly places the angular position of the planet in the proper quadrant and does not fail when $x = 0$. This is exactly how the initial azimuth is calculated in CENTF.wk1, but it will not do for the subsequent calculations of ϕ. The reason is that this calculation always gives a result for ϕ that lies between $+1/2$ and $-1/2$ revolution; we want ϕ to continue to increase or decrease by one revolution for each successive orbit.

Exploration

Once you have completed CENTF.wk1 (see Exercise 10.4) and checked its numerical results against the case given in Screens 10.1 and 10.2, take some time to familiarize yourself with the worksheet's ten predefined graphs. For various values of the initial conditions, view each of the graphs in turn and identify how each graph's presentation of the planet's orbital motion relates to those of the other graphs. Also study the various parameters given in the Results section of the worksheet's I/O page. How do these seven values vary as you change initial conditions? How do they depend on one another? How do their

```
O137: PR -X*|F/R|                                                    READY
```

	L	M	N	O	P	Q	R	S
	time	pos'n	vel_n+1/2	accel	vel_n	pos'n	vel_n+1/2	accel
135	t	x	vxhs	ax	vx	y	vyhs	ay
136								
137	0.000	1.100	-0.245	-32.627	0.000	0.000	5.674	0.000
138	0.015	1.096	-0.733	-32.551	-0.489	0.085	5.636	-2.527
139	0.030	1.085	-1.218	-32.323	-0.975	0.170	5.560	-5.052
140	0.045	1.067	-1.697	-31.940	-1.457	0.253	5.446	-7.574
141	0.060	1.042	-2.168	-31.399	-1.932	0.335	5.295	-10.091
142	0.075	1.009	-2.628	-30.696	-2.398	0.414	5.106	-12.598

```
V137: PR +$GM/R^$|N+1|                                               READY
```

	T	U	V	W	X	Y	Z	AA
	vel_n	radius	scratch	energy	ang mom	rad'l mom	speed	azimuth
135	vy	r	|F/r|	E	L	p_r	V	phi
136								
137	5.674	1.100	29.661	-19.8238	6.241087	0.000	5.674	0.000
138	5.655	1.100	29.691	-19.8239	6.241087	-0.050	5.676	0.012
139	5.598	1.099	29.781	-19.8241	6.241087	-0.099	5.682	0.025
140	5.503	1.097	29.932	-19.8243	6.241087	-0.148	5.693	0.037
141	5.371	1.094	30.145	-19.8247	6.241087	-0.197	5.708	0.050
142	5.201	1.091	30.419	-19.8252	6.241087	-0.244	5.727	0.062

Screen 10.2: The first few rows of the work block of the CENTF worksheet with values matching those in Screen 10.1.

values correlate with the changing appearance of the various graphs?

For noncircular orbits, r varies periodically. While CENTF.wk1 uses rather fancy database statistical functions to evaluate the repetition period of r as function of each time and azimuth, a careful inspection of the two graphs RV_T and R_PHI should help you grasp the physical significance of these two distinct periods. Before going on to do the exercises, think about what the orbit of a planet would look like if the azimuthal period of r were slightly larger or smaller than 1 revolution. (A function $r(\phi)$ satisfying $r(\phi) = r(\phi + 1 + \Delta\phi)$ has an azimuthal (ϕ) period of $1 + \Delta\phi$ revolutions.) Sketch your prediction for the shapes of such orbits and keep the sketches to compare with the orbits you observe as you do the exercises.

With time steps dt small enough to give reasonable accuracy, you may often fail to see an orbit continue as long as you wish on a single graph. A partial cure for this, allowing the subsequent motion to be seen on a subsequent graph, is to copy the final positions and velocities at the bottom of the work block into the initial data cells on the I/O page. This should be done using the Range Value command (see Chapter 3, page 54). In the three distributed worksheets for this chapter this operation is automated by a **macro**—a remembered sequence of keystrokes—that can be executed simply by pressing {Alt-M}. These and other macros are described in more detail in Appendix C.

EXERCISE 10.1^C Show that equations 10.1 correctly represent the Cartesian components of the force exerted on m by M. In particular, verify that the minus sign in each equation is needed to represent an attractive force acting on m.

EXERCISE 10.2^P Show that the numerical value, for the product of the gravitational constant G and the mass of the sun M in the system of units defined in this section, is $GM = 4\pi^2 \, (\text{AU})^3/\text{yr}^2$. Remember that the value of GM does not depend on the details of the planet's orbit, so it can be evaluated by applying what you know about uniform circular motion to a planet with a circular orbit of radius 1 AU and orbital period of 1 yr. If the mean radius of Jupiter's orbit had been chosen instead for the length unit, what time unit would be needed to make GM still have the numerical value $4\pi^2$?

EXERCISE 10.3^E Starting from the basic definition of the angular momentum L, show that L/r^2 is the time rate of change of the angular position of the planet (for $m = 1$). Using this result, write a short justification for the form of the expression for ϕ given in the set of equations 10.2. As part of your answer, name the numerical technique used to solve for successive values of ϕ.

EXERCISE 10.4^C Complete the CENTF worksheet using the set of equations 10.2. You may find that you want to disable global protection to do this. Remember to enable it again after you have finished editing the worksheet. Screens 10.1 and 10.2 show a set of system parameters from the home block and the first few rows of the work block corresponding to them from the completed worksheet. When your completed version of CENTF agrees with the sample, print out a listing of the equations you have used in the first two rows of computation in the work block. If you want to save your modifications in the same compacted form in which CENTF was distributed, see Appendix C, Section C.1 for the use of the {Alt-S} macro, which does this job.

EXERCISE 10.5^P For a planet whose azimuthal r-period is not exactly one revolution, its temporal r-period is different from the time it takes to make one revolution. Which, if either, of these two different time periods corresponds to one year on this planet. Justify your answer in terms of how you would define a year.

EXERCISE 10.6^C Using CENTF.wk1 with the values of the construction parameters and the initial conditions specified in Screen 10.1, explore the effect of changing the step size dt. Print two graphs: one to show the orbits as closed ellipses by use of an adequately small value of dt, the other to show moderate energy errors when dt is larger. Be sure to write on each graph (or include in its title) the value of the step size dt that is used to produce it.

EXERCISE 10.7^E Repeat the previous exercise for more elliptical orbits. You can construct such orbits by separately varying both the initial position and the initial velocity of the planet. Notice that the largest variations in the total energy occur where the value of (F/r) is largest. Choose some reasonable criterion for an acceptably small fractional variation in the total energy (say perhaps, 2%), then develop a rule of thumb for selecting a value of the time step dt that keeps the observed variation in the total energy below the variation you have chosen. Express your rule in terms of the maximum value of $|F/r|$ during this part of the orbit. Test your rule by using the {Alt-M} macro to follow an elliptical orbit around the sun for a few revolutions, changing the value of dt according to your rule before each continuation of the orbit. If you set dt too large (or unnecessarily small) during any one calculation, you should adjust the value of dt and repeat that part of the orbit with the {Calc} key, {F9}, until your observed energy variation is acceptably small, before doing your next {Alt-M}.

EXERCISE 10.8^C For an initial position of $(x, y) = (1, 0)$, find the value of the initial velocity, in units of 2π, that gives a circular orbit. Repeat for initial positions $(2, 0)$ and $(3, 0)$, adjusting dt to keep the variation in E reasonable. Formulate a rule for constructing circular orbits of arbitrary size and test its validity for other values of the initial position. Is your rule consistent with what you know about the radial force required to keep an object moving in a circle at a constant speed?

EXERCISE 10.9^C Starting with an initial position of $(x, y) = (1, 0)$ and an initial velocity of $(v_x, v_y)/2\pi = (0, 1)$, decrease v_y gradually and note its effect on the eccentricity of the orbit, being careful as usual to reduce dt as your planet approaches the sun. Can you construct a formula relating the eccentricity to $(v_y/2\pi)$?

EXERCISE 10.10^P Using CENTF.wk1 as your guide, explain how you would construct a family of bound state $(E < 0)$ orbits having the same total energy but different eccentricities.

EXERCISE 10.11^P Starting from a circular orbit of radius 1 AU, gradually increase your planet's initial velocity and note the effect this has on the geometry and the total energy of the resulting orbit. As the orbits get larger you will need to repeat {Alt-M} to inspect the full trajectory. Increase dt as you move away from the sun, but don't forget to decrease it when you draw near the sun again. Watch that the total energy E remains constant to within a few percent for each orbit and note how orbits having $E < 0$ differ from those with $E > 0$. What happens when $E = 0$?

EXERCISE 10.12^C Confirm that Kepler's third law is valid for at least six distinct orbits calculated using CENTF.wk1. For cases where a single calculation of the worksheet includes the actual minimum and maximum values of r, take the value of r_mid on the I/O page as the length of the semimajor axis. For orbits that require one or more {Alt-M}s to close, you will need to determine the orbital period—and perhaps the value of r_mid—by hand.

EXERCISE 10.13^P Starting with a noticeably elliptical orbit and holding your initial condition fixed, observe and describe what happens to orbits with force-law exponents that differ slightly from $n = 2$. Comment on both the visual appearance of the orbits and the changes in the output parameters from CENTF's I/O page.

EXERCISE 10.14^E Repeat the previous exercise for values of n ranging from -2.5 to $+2.5$ in steps of $+0.5$. What general conclusions can you draw from the observed behavior of these orbits? The case of $n = -1$ is just a linear restoring force, corresponding to a mass on an ideal spring moving in two dimensions. In this case does the mass move as you would have predicted?

10.2 ♠ RSCAT: Rutherford Scattering

As you probably noted while experimenting with CENTF.wk1, closed orbits always have negative total energy. Such orbits are termed **bound states** because the planet does not have sufficient kinetic energy to overcome the attractive potential in which it finds itself. Its distance from the sun is limited to at most that value of r at which its kinetic energy is zero because its potential energy is equal to its total energy. On the other side of $E = 0$, for positive values of the total energy, orbits are open in the sense that the object approaches the sun, has a relatively brief encounter with it, then leaves the vicinity of the sun never to return again. Since the object is moving in a straight line when it is far from its encounter with the sun and again moves along a different straight line long after its encounter, the only effect of this single encounter is to deflect the object from its original trajectory through some angle θ_{scat} to a new trajectory. For this reason, such open orbits are termed **scattering states**.

Our goal in this section is be to discover the circumstances under which an incoming projectile will be scattered through a relatively large angle, say $\theta_{scat} \geq 90°$. The case you will study models the scattering of an energetic comet by the sun, or of a spacecraft by a planet in a close flyby encounter. But since both Newtonian gravity and the Coulomb electrostatic force are $1/r^2$ force laws, these gravitational trajectories are closely related to the historic experiment in which Rutherford demonstrated the small size of the atomic nucleus by scattering alpha particles off gold atoms. One of the exercises at the end of this section invites you to explore this connection more fully.

While you could carry out your investigation with `CENTF.wk1`, a revised version of `CENTF` named `RSCAT.wk1` is provided in the student software package. `RSCAT` is complete except for the formulae for a_x and a_y which you must provide; just use the expressions for these acceleration components from equations 10.2. Its work block is identical to that of `CENTF`; its I/O page, notes page, and collection of macros have been extensively modified to ease your study of the scattering states of this system. Screen 10.3 shows its home block.

What Is the Question?

An aphorism in physics is that a problem is half solved once the question is properly stated. What would be useful to know about scattering orbits? Roughly, we want to know how hard it is to hit the target. Of all the debris floating about beyond the solar system, which pieces will act as nonperiodic comets and get disturbed by an encounter with the sun? which will not? Or, as the next section considers, how well must a spacecraft be aimed toward Mars in order to be on a very different orbit after its encounter with Mars? Figure 10.1 displays a plausible scattering orbit and gives the parameters we need to describe it. The scattering center GM is at the origin, and a particle of mass m is launched from far away with velocity v_∞. After coming in to a minimum distance r_0 it moves away in a different direction, eventually returning to the same speed v_∞ with which it began (conservation of energy, $U(\infty) = 0$). It has changed direction by an angle θ_{scat} in the process.

The least obvious parameter in the sketch is b, which is called the **impact parameter**. It describes how well the incident particle was aimed and is the

```
D16: (F0) PR [W8] @ATAN(VY_LAST/VX_LAST)*180/@PI                    READY
```

```
          A        B        C        D        E        F        G        H
1    your name  &  today's date
2    your class &  section          "Rutherford" scattering
3                                    Leapfrog integration
4    ----------------------------------------------------------------------
5    Construction parameters:                        #2pi = 2*@PI
6            Central Mass  GM =  +#2PI^2              n-1 =        1
7            Force exponent n =     2.0               n+1 =        3
8                  F = -GM/r^n
9    Initial conditions:              t_init =  0.000 Energy =  10.000
10           x-Position:    0.000    x-Veloc:  9.948 AngMom =   9.948
11           y-Position:   -1.000    y-Veloc:  0.000      b =    2.224
12                           control parameter:  0.563 = b(E/GMm)
13   Approximation:
14             dt =    0.001
15   Results (at current endpoint):
16           final direction:        28  degrees
17           scattering angle:       57  degrees
18   ----------------------------------------------------------------------
19   Notes page:        Use Alt-M for More of the current orbit
20                      Use Alt-I to Initialize a new orbit (except Energy)
     RSCAT.WK1
```

Screen 10.3: The home screen of the RSCAT worksheet.

distance at which a straight line trajectory will pass the scattering center. Fortunately it is easily computed from the quantities we use in the work block calculations—it is just the "moment arm" measuring the distance to the origin from the line of action of the initial velocity vector $\mathbf{v}_\infty$. Equivalently it is the magnitude of $\mathbf{r}_\perp$ in the formula $\mathbf{L}/m = \mathbf{r} \times \mathbf{v} = \mathbf{r}_\perp \times \mathbf{v}$ at early or late times when $|\mathbf{v}| = v_\infty$. We compute it then from $L/m = bv_\infty$

In terms of these descriptive parameters, the goal of a study of scattering orbits can be to produce a graph, for some fixed energy $E/m = \frac{1}{2}v_\infty^2$, showing how the scattering angle θ_{scat} depends on the accuracy of our aim b. We also want to know how this graph changes as we change the energy. Using dimensional analysis allows us to answer both these questions with a single graph.

Dimensional Analysis

The Newton's law equation we are solving, a vector form of $a = -GM/r^2$, contains only a single parameter GM. How, from this, can we make a graph whose x-axis is a dimensionless representative of the impact parameter b? The dimensions of GM are most easily seen from the energy formula $E/m = \frac{1}{2}v^2 - GM/r$. Each term in this formula must have the same dimensions, so the ratio is a dimensionless $rv^2/2GM$ and GM has dimensions length^3/time^2. It does not give any natural length built into the mechanism itself. Such

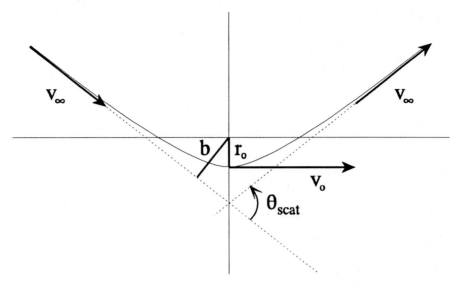

Figure 10.1: A scattering orbit. Our computations begin at $\mathbf{r}_0 = (0, -r_0)$ with an initial velocity $\mathbf{v}_0 = (v_0, 0)$ computed from the specified energy E/m and follow only the right half of the orbit. The scattering angle θ_{scat} is twice the final direction of $\mathbf{v}$ measured counterclockwise from the x-axis. The impact parameter b gives, from $L/m = bv_\infty$, the total angular momentum for the orbit and also describes how close to the target mass (GM at the origin) the projectile was originally ($t = -\infty$) aimed.

a length must therefore be constructed from the initial conditions; then the dimensionless ratio $bv_\infty^2/2GM = bE/GMm$ provides a ratio of b to the length $2GM/v_\infty^2 = GMm/E$. We therefore find that a single graph of scattering angle versus the control parameter bE/GMm serves, for every energy E, to describe how the scattering angle depends on b. Spending three minutes to follow (with worksheet calculations) a single orbit to find its scattering angle and then repeating this twenty times to produce a detailed graph of θ_{scat} versus b is an hour's hard work—you can appreciate that an hour studying dimensional analysis will save hours of preparing a dozen similar graphs for a variety of different energies!

Using RSCAT

Before you start using RSCAT there are a number of important points to understand. Since the scattering orbit is symmetric about the point where the projectile is closest to the sun (its **perihelion**), only half of each scattering orbit is calculated. The projectile is simply placed at the closest approach distance r_0 from the scattering center and given the right initial velocity v_0 to have the specified total energy E/m. By choosing this perihelion distance as our unit of length (instead of the AU as in `CENTF`), the initial conditions are always $y_0 = -1$, $x_0 = 0$, and $v_{y0} = 0$ with v_{x0} calculated from the specified energy, which is the only adjustable initial condition. From these initial conditions, other parameters of interest are then automatically calculated, including the incident velocity $v_\infty = \sqrt{2E/m}$, the angular momentum $L/m = r_0v_0$, the impact parameter $b = (L/m)/v_\infty$, and the dimensionless control parameter $b(E/m)/GM$, which is needed for the main summary graph. A macro, {Alt-I}, is provided that resets all these initial conditions for whatever energy you have specified, leaving the worksheet ready for you to press {Calc} to begin a new scattering-state trajectory.

Gravitation is a long range force in the sense that the sun will continue to bend the projectile's trajectory out to distances many times the distance of closest approach of the projectile to the sun or other scattering center. You will need to use {Alt-M} repeatedly to continue the calculation of the projectile's trajectory out until its scattering angle stabilizes. As the projectile moves farther out, the time step should be increased (typically by a factor of 5 or 10) whenever the energy errors get ridiculously small. You can monitor your progress by making the current graph one of the fixed-scale Y_X graphs or the energy E_T graph. At the end of each sequence of {Alt-M}s you can use the {Alt-R} macro to record automatically the scattering angle and corresponding energy and impact parameter in a table on the notes page. Then you can set the energy to a new value and use {Alt-I} to prepare to calculate a new trajectory. See Appendix C for more information on macros in general; for details about {Alt-I} you must examine its definition in the worksheet.

You may object that these initial conditions are too restrictive. Can we get away with changing the length scale (to match the distance of closest approach) on every different orbit and still compare these orbits meaningfully? The answer is yes if we compare only dimensionless variables, such as the scattering angle versus dimensionless impact parameter bE/GMm on the principal graph. Exercise 10.2 showed that if the unit of length is changed, a correlated change in the time unit still allows us to compute with $GM = 4\pi^2$.

EXERCISE 10.15[C]   By varying the energy and noting the resulting scattering angle, determine the value of $b(E/GMm)$ which results in 90° scattering of an object by the sun. Being careful of your units, calculate from this result the impact parameter, in AU, for an object whose incident velocity v_∞ is twice the earth's orbital velocity. Since any scattering event yielding a scattering angle greater than 90° must have an impact parameter smaller than this value of b, the impact parameter disk associated with such backward scattering has an area πb^2. Calculate this area for your value of b and express it as a fraction of the area enclosed within the earth's orbit.

EXERCISE 10.16[H]   A calculation very similar to the one in the previous exercise led Rutherford to conclude that each atom had a compact, massive nucleus at its center. In his experiment, the incident projectiles were energetic alpha particles and the role of the sun was played by gold atoms. The central force was Coulomb repulsion. Read about Rutherford's experiment in a modern physics text, revise the force law and initial conditions and rethink the units in `RSCAT`, and repeat Rutherford's calculation of the cross section for back scattering of MeV alpha particles on gold. Use your numerical result to support Rutherford's conclusion that the atom's positive charge and mass are both concentrated in a small region within each gold atom.

EXERCISE 10.17^E Non-inverse-square-law forces offer some interesting examples of unusual scattering behavior, so explore the case of $n \neq 2$. An interesting starting point is $n = 2.8$ and $E = 0.1$, in which the projectile completes nearly two full revolutions during its encounter with the sun. Can you find cases that make even more loop-the-loops around the scattering center before emerging? Watch your choice of time step here to be sure your results are not due to numerical errors.

EXERCISE 10.18^E Suppose that you want to apply these calculations to a space-craft encounter with Mars. Assume the length unit (distance of closest approach) is the diameter of the planet Mars and set $GM_{\text{Mars}} = 4\pi^2$. What then is the time unit being used in the work block, expressed in hours? (Hint: what is the period of a satellite in a circular orbit about Mars at an altitude of one Martian radius? Answer this in both conventional and work block units.)

EXERCISE 10.19^E Complete the SCATTERING graph that is laid out in the RSCAT worksheet by collecting data (using the {Alt-R} macro) for several completed scattering orbits using different initial conditions.

10.3   ♠ SLING: Slingshot Orbit for a Spacecraft

The problem modeled in this more advanced investigation is a simplified version of the recently completed Voyager mission to the outer planets. The trick here is to select the timing and initial velocity of a spacecraft launched from earth so that it gains a significant amount of energy from a close encounter with an outer planet, propelling the spacecraft still further out without using rocket fuel to provide the additional energy needed.

The actual Voyager missions, in which Jupiter provided the gravitational boost, are too computation intensive for a spreadsheet, but we can model them by changing a few parameters. One major change is to get our gravitational boost from Mars instead of Jupiter, which makes our target closer and easier to reach, even computationally. The second major change is to make Mars a much larger target than it really is. This is done by giving Mars a mass of 10^{-3} solar masses instead of the real 3×10^{-7} solar masses. As a minor simplification we put earth and Mars in simple circular orbits programmed analytically, while numerical integration is only used for the spacecraft. In addition the launch from earth is regarded as a separate piece of the problem that will have been solved elsewhere. Therefore earth's gravitational field is ignored. The initial conditions for the spacecraft orbit are treated as the velocity the spacecraft achieves relative to earth when it gets beyond significant influences of earth's gravity.

The SLING worksheet template provided in the student software package is fairly complete. You must supply the forces on the spacecraft and an analytical orbit for earth. Units are years and AU as in CENTF. The initial conditions on the I/O page are used only when $t_0 = 0$. The actual initial conditions that are used in the top row of the work block come from an auxiliary calculation done on a scratch page at **I1** to the right of the home page. If $t_0 \neq 0$ the initial data are taken from continuation values that the {Alt M} macro provides.

The initial data that you specify on the I/O page are the launch velocity magnitude and direction relative to earth. (The direction is to be specified as degrees outward toward Mars from earth's velocity vector.) Consider it a measure of your skill as a space navigator to minimize your launch speed while still reaching the outer solar system. Another initial datum is the longitude of Mars relative to earth. This is the now familiar launch window idea; we must choose a time when Mars is in a favorable position relative to earth to launch the spacecraft.

The **Results** section of the I/O page records the distance of closest approach to Mars during the last computation of the work block. The energy of the spacecraft at the beginning and end of the work block are also noted. This energy includes only the spacecraft kinetic energy and the potential energy of its interaction with the sun; it is therefore not conserved during the encounter with Mars. A successful encounter should raise this energy significantly.

Don't accept easy triumphs in your quest to reach the outer solar system! Some Mars encounters that give a great boost to the spacecraft can be totally fraudulent due to numerical error during a close encounter with Mars. Near Mars the natural time scale can be as brief as several hours, rather than the year or two period for a solar orbit at the sun–Mars separation. The worksheet includes an energy test to warn of just such numerical errors. It displays the energy of the spacecraft as seen from Mars, i.e., the kinetic energy of motion relative to Mars, plus the potential energy of the spacecraft–Mars interaction only. The solar gravitational potential is omitted. To justify this omission, you can think of this as the familiar impulse approximation. Since the Mars gravitational force is very large during a short time, other forces can be neglected during that short time, with Mars as an inertial reference (constant velocity) as long as the encounter takes a small fraction of a Martian year. On a deeper level of general relativity, the approximation is even more apt, since the principle of equivalence says that the average solar gravitational field can be neglected. Only solar gravitational field gradients make the Mars–spacecraft encounter different from a two-body problem. For numerically plausible computations, then, this Voyager–Mars energy **E_VM** should be constant during the close encounter. It will not be conserved at other times when the solar gravitational field acts differently on Mars and the spacecraft.

SLING.wk1 has a number of rather sophisticated graphs defined to help you with your mission. For example, you will see data labels on some of the orbit plots to provide controllable time markers. These will allow you to verify that the Mars and Voyager orbits cross when the two objects are simultaneously near the crossing point. Another trick, which is very useful here, is to force Lotus, beyond its apparent design, to allow three different **X** ranges as well as different `Y` ranges on `XY` graphs. Doing this involves a careful placement of the work block columns (so the x columns for earth, Mars, and Voyager are adjacent) and judicious use of blank columns preceding the Mars and Voyager y columns. Interested Lotus aficionados can study the distributed worksheet.

EXERCISE 10.20[H]   Complete the `SLING` worksheet provided in the student software package and carry out a series of numerical experiments to determine the optimal launch conditions that maximize the boost the spacecraft receives from Mars while minimizing the fuel required to give the spacecraft its initial velocity. A lot of strategy is needed to find a good trajectory. A good way to start is to set Mars' mass to zero and just look for an orbit that crosses Mars' orbit without requiring a high launch velocity. (A prior step should be to check that with zero launch velocity the spacecraft follows earth's orbit.) Once a trajectory is found that provides a small closest approach distance, one can reinstate a Mars mass and try to adjust the Mars longitude at Voyager launch to get a positive energy boost from the encounter. A good energy boost means as close to 180° scattering (as seen from Mars) as possible, for maximum momentum transfer. Good luck on your mission!

Chapter 11

Oscillations

The first sections in this chapter treat the simple harmonic oscillator in ways that extend the treatment in your text. Dimensional analysis and numerical methods are emphasized, methods that are also applicable to nonlinear oscillators. The optional sections give some examples of nonlinear oscillations, where research has recently been invigorated by extensive numerical experiments.

11.1   Simple Harmonic Oscillator

The standard example leading to simple harmonic oscillations is a mass held near an equilibrium position $x = 0$ by an ideal spring that exerts a force $F = -kx$. Then Newton's second law gives

$$m\ddot{x} = -kx \tag{11.1}$$

which we rewrite, dividing by m, as

$$\ddot{x} + (k/m)x = 0 \quad . \tag{11.2}$$

Your text probably solves this equation analytically in one of the forms

$$
\begin{aligned}
x &= A\cos(\omega_0 t + \delta) \\
&= x_0 \cos \omega_0 t + (v_0/\omega_0)\sin \omega_0 t
\end{aligned}
\tag{11.3}
$$

with $\omega_0 = \sqrt{k/m}$. We will briefly solve the same problem numerically and see in the resulting graphs the variety of solutions corresponding to different choices of A and δ or of x_0 and v_0 in the analytic solution.

Dimensional Analysis

The first step in a numerical approach is dimensional analysis. This avoids multiple computations resulting in otherwise identical graphs that differ only in the scale factors shown on the axes. In equation 11.2 we see that the first term has dimensions $[L/T^2]$ or length divided by time squared. The second term must therefore have the same dimensions. Since the x supplies the length $[L]$, the coefficient k/m must have dimensions of reciprocal time squared $[1/T^2]$. The definition

$$\omega_0^2 \equiv k/m \qquad\qquad (11.4)$$

is a mnemonic for this $[1/T^2]$ dimension of k/m —physicists conventionally prefer the letter ω for quantities like angular velocities and angular frequencies that have dimension $[1/T]$. Equation 11.2 then becomes

$$\ddot{x} + \omega_0^2 x = 0 \qquad\qquad (11.5)$$

and we thus find that a characteristic time $1/\omega_0$ is built into our apparatus.

We adopt the natural (built-in) time interval $1/\omega_0$ as the unit of time to be used in the numerical computations. As a result, the pure numbers in the t column in the worksheet are the values of the dimensionless ratio $t/(1/\omega_0) = \omega_0 t$ (instead of being values of the dimensionless ratio $t/(1\,\text{s})$ when SI units are used). This is commonly expressed in two ways. We can say, "in these units, then, $\omega_0 = 1$", or we can say "let us write t for what would previously have been called $\omega_0 t$" (as we did in optional Chapter 9 at page 134). In either case, the result is that numerical computations are based on the equation

$$\ddot{x} + x = 0 \quad . \qquad\qquad (11.6)$$

Length, mass, and time are the basic units in mechanics, but we have not found a natural length unit in this ideal apparatus (m provides a mass unit when needed). You should not be surprised, then, to find that similar motions will be predicted on all length scales when you solve this equation.

Numerical Solution

Equation 11.6 was solved in the **LEAP** worksheet (Section 4.2 at page 65). Otherwise it can be solved with the frequently used template **CBOND** (Section 5.2 at page 74)—you simply modify the acceleration column to compute $a = -x$ and the energy column to compute $E/(m\omega_0^2) = \frac{1}{2}\dot{x}^2 + \frac{1}{2}x^2$. We want to explore the solutions of equation 11.6 that these worksheets can graphically present and to use a selection of such graphs to verify the following important properties of these solutions:

Period. For any initial conditions, the solution $x(t)$ repeats its motion after a fixed time T, so that $x(t) = x(t + T)$. In our units (where $1/\omega_0$ is the time unit) this period is 2π, so that in any other units it is $T = 2\pi/\omega_0$.

Amplitude. The motion is bounded with $|x| \leq A$ where the (least upper bound) amplitude A can be found from the energy as $E/(m\omega_0^2) = \frac{1}{2}A^2$.

Phase shifts. Different solutions that have the same amplitude differ only by a time translation—the graph of one such solution can be made to overlay any other simply by sliding it sideways so that the zero of time on one is shifted relative to the other.

Scale invariance. If two initial condition sets differ only in scale, the same is true of the corresponding solutions; i.e., if $x = f(t)$ is one solution, then $x = \lambda f(t)$ is also a solution for any constant scale factor λ.

Sinusoid. The numerical solutions agree (approximately) with the analytic solution in terms of sine and cosine functions from equation 11.3.

EXERCISE 11.1^{E} Verify that of the two analytic forms given in equations 11.3 each satisfies the differential equation 11.5. Show the steps you take to evaluate $\ddot{x}$ in this process.

EXERCISE 11.2^{H} Use trigonometric identities to convert the first (A, δ) analytic form in equations 11.3 to the second (x_0, v_0) and, in this way, obtain expressions for x_0, v_0 in terms of A, δ.

EXERCISE 11.3^{C} Build a worksheet that solves equation 11.5 numerically using the leapfrog method and provides graphs of the solution, including an energy graph. You should begin with a debugged copy of LEAP or CBOND. Print a phase-plane graph, and a graph or (short) table of energy versus time.

EXERCISE 11.4^{C} Print an $x(t)$ graph of equation 11.5 for a nontrivial solution (not $x = 0$) of equation 11.5 where periodicity is evident. Mark with pencil on the graph several pairs of points that are one period apart. Read off the time coordinates of these points and compute the difference (period T) for each pair.

EXERCISE 11.5^{C} Use judicious choices of the integration step dt in your worksheet to get an accurate numerical estimate of the period T of the solutions of equation 11.5. How much accuracy do you think your computations support? (State an estimated uncertainty.) What is the basis for your estimate? Repeat your best computation using a very different amplitude by changing the initial conditions. How much did the period change?

EXERCISE 11.6^{E} Print at least three graphs of $x(t)$ that give the same amplitude $\max_t x(t)$, but are produced from different initial data at $t = 0$.

EXERCISE 11.7^{E} Print graphs of $x(t)$ with $v_0 = 0$ but with $x_0 = .001, 1, 1000$ and 10^6. Describe in a sentence or two their similarities and their differences.

EXERCISE 11.8^{E} Add a column to your worksheet that computes the analytic solution from equation 11.3 for the same initial conditions that produce the numerical solution. Compare them on a single graph (use distinct formats and clear legends) and print examples where they nearly coincide and where there are clear differences.

11.2   OSC: Damped Driven Oscillations

Equation of Motion

The harmonic oscillator arises in many contexts, usually as an approximation that applies only to small oscillations about some equilibrium state of a physical system. Your text may offer a variety of examples and exercises where Newton's laws, applied to a suitable idealized system, lead from the general $m\ddot{x} = F(x, \dot{x}, t)$ to an equation of the form we will study here,

$$\ddot{x} + \dot{x}/\tau + \omega_0^2 x = \omega_0^2 G \cos \omega t \quad , \tag{11.7}$$

with different constants appearing where we have written $1/\tau$, ω_0^2, and $\omega_0^2 G$. Equation 11.2 is a special case of equation 11.7 if we identify $1/\tau = 0$, $\omega_0^2 = k/m$, and $\omega_0^2 G = 0$.

In the previous example $2\pi/\omega_0$ was found to be the period of the motion; similarly the constants in equation 11.7 acquire significance by their relationship to the behaviors of the solution, which we need to explore either analytically or numerically.

A standard example giving equation 11.7 includes a viscous drag force $F_{\text{drag}} = -b\dot{x}$ and a motor to move the end of the spring opposite the mass, so that the mass's zero-force position x_z, which was $x_z = 0$ in equation 11.1, becomes $x_z = Z \cos \omega t$. The spring force is now

$$F_{\text{spring}} = -k(x - x_z) = -k(x - Z \cos \omega t) \tag{11.8}$$

and Newton's second law gives

$$m\ddot{x} = F_{\text{spring}} + F_{\text{drag}} = -k(x - Z \cos \omega t) - b\dot{x} \quad . \tag{11.9}$$

We again divide by m to see the dimensions of each term better and to compare this equation with equation 11.7. The result, somewhat reorganized, is

$$\ddot{x} + (b/m)\dot{x} + (k/m)x = (k/m)Z \cos \omega t \quad . \tag{11.10}$$

It has the form of equation 11.7 with $1/\tau = b/m$, $\omega_0^2 = k/m$ and $G = Z$. (For another briefly outlined example refer to page 167. There you see how equation 11.7 arises in electrical circuits, in which case $1/\tau$, etc., are related to electrical quantities.)

Dimensional Analysis

The main task of this section is to see what variety of motions are allowed when equation 11.7 controls the motion and to see the significance of the constants in that equation through their influence on the motion. The first step in any such survey is to distinguish between dimensioned constants, which control the scale of the graphs we draw, and dimensionless constants, which will have scale-independent effects on the motions. In numerical solutions only the dimensionless constants need be varied; independent dimensioned constants can be incorporated as we interpret the graphs.

Equation 11.7 has been written in a form that makes the dimensions of the constants easy to see. For terms to be added together, each term must have the same dimensions. The first term $\ddot{x}$ is seen to have dimensions $[L/T^2]$, i.e., length divided by time squared. Looking at the next term, $\dot{x}/\tau$, we see that τ must have dimensions of time, since $\dot{x}$ has dimensions $[L/T]$. The third term $\omega_0^2 x$ shows, as before, that ω_0 is dimensionally a frequency or reciprocal time $[1/T]$. We have therefore found two time scales or frequencies in the equation and can define a dimensionless number as the ratio of the two time scales τ and $1/\omega_0$. This number is conventionally called Q and we define it as

$$Q \equiv \omega_0 \tau \quad . \tag{11.11}$$

When there is a driving force, represented by the term $\omega_0^2 G \cos \omega t$, we can find a length G in the problem. (If there is no driving term of this sort, so $G = 0$, a characteristic length must be sought among the initial conditions of the motion.) The driving force is assumed to have a simple sinusoidal time dependence proportional to $\cos \omega t$, which brings in another constant ω with dimensions of a frequency $[1/T]$. This we see from the fact that the cosine function requires a pure number as an argument, so ωt must be dimensionless and ω must have dimensions reciprocal to t. In numerical computations we want to use a dimensionless representative of ω, which we define as

$$\Omega \equiv \omega/\omega_0 \quad . \tag{11.12}$$

The result of the dimensional analysis is that there is always a time $1/\omega_0$ defined (if there is any linear restoring force) and there will be a length scale G defined if there is a driving term. These can be used as our basic units (the particle's mass m defines a mass unit when that is needed—it doesn't appear separately in equation 11.10). Damping (viscous drag) introduces another time scale τ that can be described by a dimensionless parameter $Q = \omega_0 \tau$. A harmonic driving force gives still another time scale $1/\omega$ that can be described by the dimensionless frequency ratio $\Omega = \omega/\omega_0$. By adopting $1/\omega_0$ as our time unit, the differential equation can be written in the form

$$\ddot{x} + \dot{x}/Q + x = G \cos \Omega t \tag{11.13}$$

which is appropriate for numerical computations. In this form t really means the dimensionless quantity $\omega_0 t$, and only the dimensionless parameters Q and Ω appear. It is also sufficient to consider only the cases $G = 0$ (no driving force) and $G = 1$ (i.e., if $G \neq 0$ take G as the unit of length). A numerical exploration therefore needs to survey one continuous parameter Q in the $G = 0$ case and two, Q and Ω, in the driven case with $G = 1$.

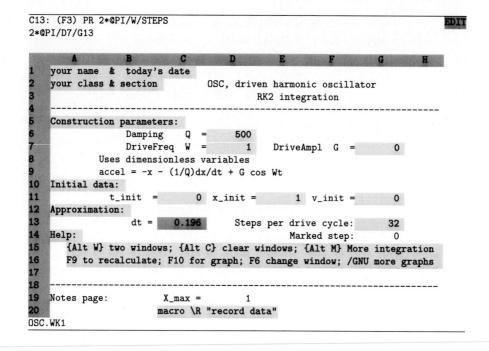

C13: (F3) PR 2*@PI/W/STEPS   **EDIT**
2*@PI/D7/G13

	A	B	C	D	E	F	G	H
1	your name & today's date							
2	your class & section			OSC, driven harmonic oscillator				
3					RK2 integration			
4	--							
5	Construction parameters:							
6		Damping Q =		500				
7		DriveFreq W =		1	DriveAmpl G =		0	
8		Uses dimensionless variables						
9		accel = -x - (1/Q)dx/dt + G cos Wt						
10	Initial data:							
11		t_init =	0	x_init =	1	v_init =	0	
12	Approximation:							
13		dt =	0.196	Steps per drive cycle:		32		
14	Help:			Marked step:	0			
15	{Alt W} two windows; {Alt C} clear windows; {Alt M} More integration							
16	F9 to recalculate; F10 for graph; F6 change window; /GNU more graphs							
17								
18	--							
19	Notes page:		X_max =	1				
20			macro \R "record data"					

OSC.WK1

Screen 11.1: The home screen of the OSC worksheet.

Numerical Solution

Our favorite leapfrog integration scheme cannot, unfortunately, be used when forces are velocity dependent, so to solve equation 11.13 we need the slightly more detailed scheme called RK2 that was described in connection with air drag forces in Section 6.4. You can either modify the RK2 worksheet described there or complete the worksheet OSC from the student software package to integrate equation 11.13. Screens 11.1 and 11.2 offer some suggestions if you modify RK2. To complete OSC all that is needed is to provide a formula for the acceleration, $a = -v/Q - x + G \cos \Omega t$, in both the whole step (**a**) column and the half step (**ahs**) column. Use the OSC worksheet if it is available, because it provides some extra conveniences for exploring the solutions.

Your first experiments with the OSC worksheet should be to check it. Try to reproduce simple harmonic oscillator behavior by turning off the driving force (set $G = 0$) and making the frictional drag small (set $Q \gg 1$, say $Q = 500$). Errors due to the RK2 approximation are bigger, however, than those in the leapfrog method, so do not expect to reproduce results accurately from Section 11.1 without using smaller time steps—32 steps per cycle of oscillation is a good starting point.

```
S143: PR @IF(@MOD(@ROUND(T/$DT,0),$STEPS)=$PHASE,"o","")              EDIT
@IF(@MOD(@ROUND($L143/$C$13,0),$G$13)=$G$14,"o","")
```

	L	M	N	O	P	Q	R	S
140		---- At time t ----			---- At time t+dt/2 ----			
141	time	pos'n	momentum	Force	pos'n	momentum	Force	mark
142	t	x	v	a	xhs	vhs	ahs	
143	0	1	0	-1	1	-0.09817	-0.99980	o
144	0.196349	0.980723	-0.19631	-0.98033	0.961450	-0.29255	-0.96086	
145	0.392699	0.923280	-0.38497	-0.92251	0.885485	-0.47554	-0.88453	
146	0.589048	0.829907	-0.55865	-0.82879	0.775061	-0.64002	-0.77378	

Screen 11.2: The start of the work block in the OSC worksheet. The content line shows the complicated formula that computes the column of graph data labels to provide strobe markers on graphs once each cycle at frequency $W = \Omega$.

Damped Oscillations

Begin exploring by watching the (numerically predicted) results of damping. Set $G = 0$ and try values of Q between 3 and 30. Look at $x(t)$ graphs, phase-plane graphs, and graphs of energy $E(t)$. Watch a real oscillator made of a mass hanging on a spring (which will always have some damping) and try to measure its Q value by producing a graph whose behavior matches that of the physical oscillator to within five or ten percent. A feature that is simple to measure is the number of oscillations (full periods) required before the amplitude of the oscillation decreases by a factor of three (or some other number). Repeat Exercise 11.7 for the present case, verify that the (undriven) damped motions are also scale invariant, and see if there is a range of (small) oscillations where your real apparatus has this property.

Small Q values $Q \leq 1$ that correspond to strong damping can give unexpected numerical problems, as well as behavior different from that in the lightly damped cases. When exploring behaviors in these cases try various dt values to be sure your values are small enough to avoid errors that change the results qualitatively.

Often damping is desirable because you want a mechanism to settle down to equilibrium as quickly as possible. With very strong damping $Q \ll 1$ the mechanism is so sticky it can hardly move; with very weak damping $Q \gg 1$ it oscillates about equilibrium forever. Some intermediate value leads to the fastest approach to equilibrium (see Exercise 11.15).

More Testing

The first verification of your worksheet—reproducing the motions of an undamped, undriven oscillator—is not very stringent in that it did not test the new features of this model. All computer programs (like other complicated engineering feats) require a lot of testing before they can be trusted to operate as intended. For programs like the OSC worksheet that solve equations numerically, a good test is to reproduce a known analytical solution. This typically requires finding a case with some special simplicity that allows an analytic solution. (The driven damped oscillator is unusual in that an analytic solution is known for the general case, but the algebra and trigonometry involved is daunting to most undergraduates.) You should be able to show that in the special case $\Omega = 1$, the formula $x = A \sin t$ solves equation 11.13 for a particular choice of the constant A. Exercise 11.17 asks you to check your worksheet against this solution.

Driven Oscillations

Let us begin studying the driven oscillator ($G \neq 0$) by trying to find yet another special case where the algebra is simple. Suppose there is no damping so equation 11.13 becomes

$$\ddot{x} + x = G \cos \Omega t \quad . \tag{11.14}$$

Then $x = A \cos \Omega t$ is a plausible guess at a solution, and we quickly see that every term in equation 11.14 then contains the common factor $\cos \Omega t$. When this factor is removed there remains just $-\Omega^2 A + A = G$ with the solution

$$A = G/(1 - \Omega^2) \quad . \tag{11.15}$$

Sketch A/G as function of $\Omega \equiv \omega/\omega_0$ or plot it (Exercise 11.16); it represents the amplitude response A at various frequencies to a driving force of magnitude G when damping is unimportant. The negative sign in equation 11.15 at high frequencies, $A \propto -G/\Omega^2$, means that the response is in the opposite direction to the force; i.e., a force $G \cos \Omega t$ produces a motion $x \simeq -(G/\Omega^2) \cos \Omega t$. You should verify this qualitatively with a long period pendulum or spring oscillator held and driven by small motions of your hand.

A numerical exploration of other cases shows that this special case is actually a good approximation in many important cases. The questions about the theoretical model (equation 11.13) that remain for numerical investigation are: How does the response to a driving force depend on initial conditions? and, How does damping change the response? Regarding the first question, in the presence of damping we find that the effects of initial conditions damp out on the same time scales as those found for undriven oscillators to reach equilibrium and that a steady-state response to the driving force emerges if the numerical solution is continued for a long enough time. How the amplitude of this steady-state solution depends on the driving frequency, the **resonance curve**, is important for understanding the behavior of driven oscillators and reflects modifications of equation 11.15 in the presence of damping. The most important modifications occur near $\Omega = 1$, i.e., near $\omega = \omega_0$ where the resulting amplitudes are large and the oscillator is driven near its natural undamped frequency.

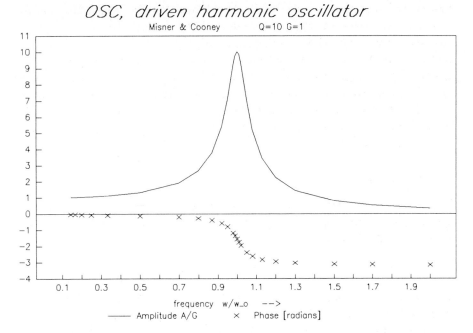

Figure 11.1: Resonance curves showing the amplitude and phase of the steady-state motion $x = A\cos(\omega t + \delta)$ produced by a driving force $F = m\omega_0^2 G\cos\omega t$. The plotted quantities are the phase δ and the dimensionless ratio A/G against $\Omega = \omega/\omega_0$. For these curves the damping was specified by $Q \equiv \omega_0\tau = 10$.

Figure 11.1 is an example of a resonance curve for the case $Q = 10$. Exercise 11.20 shows how the data for such a graph can be collected numerically. These numerical methods also work for nonlinear oscillators that reach a steady-state (periodic) oscillation after damped transient behaviors die out. For the harmonic oscillator an analytic solution in the form $x = A\cos(\omega t + \delta)$ can be found by experienced and skillful use of trigonometry or complex number algebra. To report our numerical explorations in this standard form we can calculate that $A = \max_t x(t)$. To find a formula yielding δ from the numerical results, note that $x(0) = A\cos\delta$ and $v(0) = -\omega A\sin\delta$ so that δ can be produced as $\delta = -\arctan[v(0)/\omega x(0)]$.

Phase Plane

When you study phase-plane $v(x)$ plots for a driven oscillator, you may be surprised to see trajectories intersect themselves, which we proved impossible on page 75. But that proof had an essential premise, namely that the force law $F = F(x, v)$ did not have any time dependence. In the presence of a driving force $F = F_0\cos(\omega t)$ a given set (x_0, v_0) of initial conditions can produce one result if specified when $\omega t = 0$ when the force is pulling, and a different result if specified when $\omega t = \pi$ so the force is pushing. The uniqueness theorem for solutions of differential equations will tell us only that two trajectories cannot cross at the same phase of the driving force. We should therefore notice the special points along a trajectory that occur when $\omega t = 2\pi n$ where n is an integer. The distributed worksheet OSC shows markers placed on these points by the /Graph Options Data-Labels command. If two such marks coincide, the trajectory must be periodic, since the motion following one mark will be identical to that following the other.

For oscillator equations without time-dependent coefficients, such as the damped but undriven harmonic oscillator ($G = 0$) and the nonlinear oscillators studied in the VdPOL and RALEH worksheets in optional Section 11.3, the premises of the no crossing theorem are satisfied. The uniqueness theorem for differential equations then applies in its simple form to the xv plane (not just in the xvt space needed when there is a time-dependent driving force).

Bells and Whistles

To make numerical exploration efficient, several conveniences have be incorporated into the OSC worksheet distributed in the student software package. One is a "More integration" macro that, after you have inspected graphs produced by integrating the differential equation from a given set of initial conditions, allows you to see what happens next. Because of the computer's memory limitations (and the way spreadsheets are designed) only a modest length of time can be explored with one worksheet calculation. To continue the exploration of motion beyond that point requires that one take as new initial conditions the position and velocity of the mass at the last point calculated. The many keystrokes it takes to do this have been recorded and are replayed simply by pressing the {Alt-M} key combination.

In the OSC worksheet you don't set the time step dt directly; instead you specify an integer n, the number of integration steps per drive cycle. Here drive cycle refers to the oscillations $G \cos \Omega t$ of the driving force. This force repeats its pull and push cycle every time Ωt increases by 2π, so one period of the driving mechanism is $\Delta t = 2\pi/\Omega$ and the formula used to set dt for n steps per drive cycle is $dt = (2\pi/\Omega)/n$. Even if there is no driving force ($G = 0$), you can set dt this way. In that case it is useful to think of Ω (or $\omega \equiv \Omega\omega_0$) as the frequency of a flashing stroboscopic light that illuminates the mechanism periodically. Most graphs in the OSC worksheet have markers placed on the curves at this frequency.

The distributed worksheet has a work block long enough to do 320 integration steps. It's best if you always choose the number of steps per cycle (at frequency Ω) to be a divisor of 320, such as 32, 40, 64, etc. Otherwise the strobe markers on the graphs will not be placed properly on the continuation runs.

Other conveniences in OSC are the {Alt-W} and {Alt-C} macros which will divide the screen into two Windows (or Clear it back to normal) so that the I/O block and the work block can be viewed simultaneously. When the screen is divided into two windows, the {F6} or {Window} key can be used to jump the cell pointer back and forth between the two windows. These and other macros are described in more detail in Appendix C.

Electrical Oscillations

The simple harmonic oscillator equation 11.7 appears in many areas besides mechanics. It occurs most frequently in electrical circuits, where it is practical to follow oscillation for thousands or millions of cycles in a few seconds and thus do experiments quickly. Damping arises as currents I moving through wires encounter resistance, which heats the wires. Potential energy is created when capacitors are filled with electric charge q, which is related to the current I flowing into the capacitors by $I = dq/dt$. The currents have "inertia" (want to continue flowing) because energy is stored in the magnetic fields these currents produce. The current can't stop until this energy is disposed of. (That's why you see a spark if you unplug a large motor like a vacuum cleaner while it is running.) For a simple circuit, the equation that results is

$$L\ddot{q} + R\dot{q} + q/C = V_0 \cos \omega t \qquad (11.16)$$

where V_0 is the AC voltage applied across the circuit and $I = \dot{q}$ is the current that voltage pushes through the circuit. Here L is the magnetic inductance of the circuit (usually in coils of wire, as in motor windings) defined so that $\frac{1}{2}LI^2$ is the magnetic energy; C is the capacitance with $\frac{1}{2}q^2/C$ the electric potential energy, and R is the resistance of the circuit so that I^2R is the rate of electrical energy loss (Ohmic heating of the wires). Divide this equation by L to be able to recognize the units of various quantities:

$$\ddot{q} + (R/L)\dot{q} + q/LC = (V_0/L) \cos \omega t \quad . \qquad (11.17)$$

From this we see that R/L has dimensions of a reciprocal time $1/\tau$, and that $1/LC$ has dimensions of reciprocal time squared, or frequency squared $1/LC = \omega_0^2$. We write $Q = \omega_0\tau = \omega_0 L/R$, $\Omega = \omega/\omega_0$ and $G = V_0/L\omega_0^2 = V_0/C$ to reduce this equation to the one studied in our worksheet.

EXERCISE 11.9[C]   Complete the OSC.wk1 file from the student software package by adding the computations needed to let it solve equation 11.13. Test it to see that it can approximately reproduce simple harmonic motion by setting $G = 0$ and $Q = 500$; use 32 steps per cycle and set markers at frequency $\Omega = 1$. Print graphs of $x(t)$ and $v(x)$.

EXERCISE 11.10[P]   Adapt the RK2 worksheet from Section 6.4 at page 81 to solve equation 11.13 and test it as described in the previous problem. Save it under a new name, such as OSCR.wk1.

EXERCISE 11.11[C]   Use your completed worksheet to follow the motion of (our theoretical model of) a damped oscillator without a driving force (set $G = 0$). Use initial conditions $x(0) = 1$ and $\dot{x}(0) = 0$ and pick a damping parameter Q in the range 3 to 10. Print graphs of $x(t)$, $v(t)$, and $E(t)$ where E is the total energy. From these graphs evaluate the number a_1 in the equation $\max_{t>0} v(t) = a_1\omega_0 x_0$ for your chosen value of Q and these $v_0 = 0$ initial conditions.

EXERCISE 11.12^{P} Do the previous exercise, but for initial conditions $x(0) = 0$, $\dot{x}(0) = 1$, to find from graphs the number b_1 in the equation $\max_{t>0} x(t) = b_1 v_0/\omega_0$ for your chosen value of Q and these $x_0 = 0$ initial conditions.

EXERCISE 11.13^{E} Repeat Exercise 11.7, but now for a damped oscillator without a driving force. Use a Q in the range 3 to 10.

EXERCISE 11.14^{E} **Overdamped** motion occurs when $Q \ll 1$. Explore solutions with $Q = 0.05$ and $G = 0$. Print graphs of $v(x)$ (phase plane) and of $E(t)$. Be sure to check that your physical results are not sensitive to dt.

EXERCISE 11.15^{E} Set initial conditions $x_0 = 1$, $v_0 = 0$ and make, on your notes page, tables of energy and of $|x|$ at time $\omega_0 t = 2\pi$ for a variety of Q values. Print a graph of these data. Locate the Q value that causes the most rapid approach to equilibrium. Which criterion, energy or $|x|$, is the best guide in this context? Why?

EXERCISE 11.16^{P} Plot $A(\Omega)/G$ from equation 11.15 using, for example, the graph template worksheet GTPLT from appendix Section D.2 at page 213. Because the function diverges at $\Omega = 1$, you may find it helpful to use manual scaling on the ordinate of your graph, /Graph Option Scale Y_scale Manual, with upper and lower limits of ± 5 or ± 10.

EXERCISE 11.17^{E} For $\Omega = 1$ solve equation 11.13 analytically in the form $x = A\sin t$ and modify your OSC worksheet to print graphs comparing this analytic solution with the worksheet's numerical solution in two cases: $Q = 2$ and $Q = 10$.

EXERCISE 11.18^{E} How do initial conditions enter the motion of a driven oscillator? In the special case of equation 11.14 where we found one particular analytic solution, show that $x = [G/(1 - \Omega^2)]\cos\Omega t + A_0\cos(t + \delta)$ also satisfies the equation. This solution contains two constants A_0 and δ that can be used to fit initial conditions. Use the OSC worksheet (at some high value of Q) and print an example of a phase-plane plot where you can see that the second A_0 term in the analytic solution thus approximated must be playing a significant role. Use a rational (simple fraction) value for Ω to get interesting patterns. Because real oscillators are usually significantly damped, such motions seldom arise in practice.

EXERCISE 11.19^{C} Find the steady-state solution when a moderately damped harmonic oscillator with $Q = 5$ is driven at an off-resonance frequency $\Omega = 1.5$ with $G \neq 0$: Print two phase-plane graphs using quite different initial conditions but the same driving force to show that after several cycles the same motion results. Also copy the x and v values at the end of several complete cycles (where $\Omega t/2\pi$ is an integer) to the initial values (using the /Range Value command introduced on page 54 or the {Alt-M} macro in the OSC worksheet) to continue the integration with smaller time steps and find and report accurate values of x/G and $v/\omega_0 G$ at the beginning of each cycle (when $\cos\omega t = 1$) in the steady-state solution, which appears after the transient effects of the initial conditions have died out.

EXERCISE 11.20[E]   For three frequencies Ω different from the 1.5 of the previous exercise, find the initial conditions that produce a purely periodic solution (steady state). Make a table on your notes page that records Ω, $\max_t x(t)$, x_0/G, and $v_0/\omega_0 G$ for these cases and print it. As in the previous problem, these desired initial conditions can be found by starting with arbitrary initial conditions and integrating long enough (use {Alt-M}) for the effects of your initial conditions to damp out. In the OSC worksheet the table headings are laid out, and the {Alt-R} macro will copy the desired data once you have continued the integration long enough.

11.3   ♠ VDPOL, RALEH: Limit Cycles in Self-Excited Oscillators

A self-excited oscillator is one in which oscillations build up without any driving force at the oscillation frequency. There must, of course, be some power source since energy is conserved in nature even if not in our models of small mechanisms. A familiar example is a bowed violin string. The bow can move across the string at a low frequency (constant velocity, $\omega = 0$) yet the string will begin to vibrate near its resonant frequency ω_0. A similar effect occurs as a result of feedback in electronic circuits. If a microphone picks up enough sound from the loudspeaker to which it is connected, then any small sound heard by the microphone comes back almost immediately as a louder sound heard by the same microphone, and a squeal soon builds up, powered by the amplifier that connects the microphone to the loudspeaker.

Linear Antidamping

The beginnings of the feedback process can be modeled by linear antidamping:

$$\ddot{x} - \dot{x}/\tau + \omega_0^2 x = 0 \quad , \tag{11.18}$$

or, in dimensionless form ($\omega_0 t$ becomes t, $Q \equiv \omega_0 \tau$),

$$\ddot{x} - \dot{x}/Q + x = 0 \quad . \tag{11.19}$$

In a mechanical model, a force $F_{\text{ampl}} = +(m/\tau)\dot{x}$ which, unlike friction, pushes in the same direction as the velocity gives rise to the $\dot{x}$ term in these equations. You can quickly see its effects just by trying negative values of Q in the OSC worksheet.[1] (Note that a negative Q in equation 11.13 has the same effect as a positive Q in the antidamped equation 11.19.) From this linear equation the amplitude increases forever, but in a physical mechanism something will happen to stop it. The van der Pol and Raleigh models studied in this section introduce limitations in slightly different ways.

---

[1] There is also an analytic energy argument that shows how oscillations build up: Corresponding to equation 11.19 there is a dimensionless energy $E = \frac{1}{2}(\dot{x}^2 + x^2)$. Its rate of change is $dE/dt = \dot{x}(\ddot{x} + x) = +\dot{x}^2/Q \geq 0$. The last equality here uses equation 11.19 to eliminate $\ddot{x}$. On average, each term in the energy formula $E = \frac{1}{2}(\dot{x}^2 + x^2)$ contributes equally, so when averaged over several oscillations the energy change formula becomes $dE/dt = +E/Q$ with solution $E \propto \exp(+t/Q) \mapsto \exp(+t/\tau)$.

The van der Pol Oscillator

The van der Pol equation in standard units reads

$$\ddot{x} = -\omega_0^2 x + \mu(\ell_0^2 - x^2)\dot{x} \tag{11.20}$$

where μ is a positive constant. In dimensionless variables ($\omega_0 t \mapsto t$, $x/\ell_0 \mapsto x$, i.e., set $\omega_0 = 1$ and $\ell_0 = 1$), the equation becomes

$$\ddot{x} - \dot{x}(1 - x^2)/Q + x = 0 \tag{11.21}$$

where

$$Q \equiv \omega_0/\mu\ell_0^2 \quad . \tag{11.22}$$

When $x^2 \ll \ell_0^2$ so that the $x^2\dot{x}$ term can be neglected, these equations just reduce to the linear antidamped equations 11.18 or 11.19 with $\mu\ell_0^2 = 1/\tau$. Therefore, initial conditions near $(x_0, v_0) = (0, 0)$ build up in amplitude as they do in linear antidamping until the neglect of this term can no longer be justified. In the opposite case, where $x^2 \gg \ell_0^2$, the sign of the $\dot{x}$ term corresponds to damping, and the energy decreases. It seems plausible, therefore, that for amplitudes of oscillation in the neighborhood of ℓ_0 there can be a periodic oscillation. Your numerical exploration with the **VdPOL** worksheet can test these hypotheses numerically, search for periodic oscillations, and see how the behaviors depend on the single dimensionless parameter Q. (Analytic methods valid for large Q are described, for example, by Ralph Baierlein in his text *Newtonian Dynamics*, McGraw-Hill, 1983.)

The VdPOL Worksheet

The van der Pol equation 11.21 can be solved numerically by minor changes to the **OSC** worksheet or to any other worksheet that you prepared to solve the damped harmonic oscillator. The acceleration formulae need to be changed in the (RK2) integration steps, and the input cell for a driving force strength G can be erased. The **VdPOL** worksheet as distributed has made one other change that can be seen in Screen 11.3: the strobe marker placement is not specified by a frequency Ω, but by the corresponding period. We specify the marking interval as a multiple of 2π in cell **G14**, so that a value 1.5 there results in markers at intervals of $\Delta t = 3\pi$. In use, we set the marker interval by trial and error to match the period of a stable oscillation when such is found. In this way we measure its period while inspecting phase-plane graphs. If the markers fall on top of each other, then they occur at exactly a period of the oscillation. An {Alt-R} macro allows you to record this period along with the corresponding Q and oscillation amplitude on the notes page of the worksheet, from which graphs of period and amplitude as function of Q can be prepared.

The Raleigh Oscillator

In the van der Pol equation, antidamping at small amplitudes is changed to damping at large x by the factor $1 - (x/\ell_0)^2$ in the $\dot{x}$ term. In a similar equation

```
S463: U @IF(@MOD(@ROUND(T_LAST/$DT,0),$STEPS)=0,"+","")          EDIT
@IF(@MOD(@ROUND($L463/$C$13,0),$G$13)=0,"+","")
```

```
            A        B        C        D        E       F       G       H
 5  Construction parameters:
 6               Damping    Q  =        2
 7
 8            Uses dimensionless units
 9            accel = -x + (1/Q)(1 - x^2) dx/dt
10  Initial data:
11            t_init  =      0  x_init =    0.1  v_init =        0
12  Approximation:
13               dt =    0.157      Steps per interval:          40
14  Help:                          Marking interval/2pi:          1
15    {Alt W} two windows; {Alt C} clear windows; {Alt M} More integration
16    F9 to recalculate; F10 for graph; F6 change window; /GNU more graphs
            L        M        N        O        P       Q       R       S
141  time    pos'n  momentum  Force    pos'n  momentum  Force    mark
142    t       x       v        a       xhs     vhs      ahs
143      0     0.1      0      -0.1      0.1 -0.00785 -0.10388   +
460 49.79424 1.561352 1.589696 -2.70420 1.686206 1.377308 -2.95559
461 49.95132 1.777699 1.125432 -2.99328 1.866090 0.890339 -2.97113
462 50.10840 1.917553 0.658727 -2.79926 1.969290 0.438873 -2.60085
463 50.26548 1.986491 0.250186 -2.35503 2.006141 0.065222 -2.10477   +
VDPOL.WK1
```

Screen 11.3: The **VdPOL** worksheet, shown divided into two windows by the use of the {Alt-W} macro.

studied by Raleigh in the nineteenth century, a factor $1 - (\dot{x}/\omega_0\ell_0)^2$ multiplying $\dot{x}$ produces damping when $\dot{x}$ is large. Although a **RALEH** worksheet has been distributed, it differs from **VdPOL** only in this change in the acceleration formula. The behavior of the two equations is quite similar when Q is large. For $Q > 1$ it requires sharp eyes to see the differences between solutions of the van der Pol equation 11.21 and the similar Raleigh equation

$$\ddot{x} - \dot{x}(1 - \dot{x}^2)/Q + x = 0 \quad . \tag{11.23}$$

In terms of dimensioned variables, this equation reads

$$\ddot{x} = -\omega_0^2 x + \nu(v_{\text{crit}}^2 - \dot{x}^2)\dot{x} \tag{11.24}$$

and involves a critical velocity above which the antidamping is reversed.

Limit Cycles

As you explore solutions of these equations using the **VdPOL** or **RALEH** worksheets (or equivalents), you should find that, as in the driven damped harmonic oscillator, the effects of any choice of initial conditions die out after some time and a steady-state (periodic) oscillation ensues. This steady-state solution is called a limit cycle in mathematics; a **limit cycle** is defined as an isolated periodic trajectory. **Isolated** means that there are no other periodic trajectories nearby in the phase plane (unlike the undamped simple harmonic oscillator where there is a periodic trajectory through every point). In these equations this limit cycle is also **stable** in the sense that nearby trajectories (which are not exactly periodic) evolve toward the limit cycle.

The origin of the phase plane is also a degenerate case of a limit cycle. The solution $x(t) = 0$ is exact and periodic (with any period you like), but it is, as we have seen, unstable. Nearby trajectories evolve away from it.

EXERCISE 11.21^{C} By inserting new columns in the work block of the OSC worksheet (or your equivalent), plot and print a graph showing $\ln E(t)$ and $-t/Q + \ln E(0)$ on the same graph, distinguished by format and legends. Use $G = 0$ and a moderate negative value of Q such as -5 to verify or contest the arguments from the footnote on page 169.

EXERCISE 11.22^{C} Complete the VdPOL worksheet by supplying the needed acceleration formulae in the work block. Test it by checking that for small initial conditions and large positive and negative Q values it antidamps/damps as expected from experience with the OSC worksheet. Print a phase-plane graph that shows this behavior.

EXERCISE 11.23^{P} Complete the RALEH worksheet and test it.

EXERCISE 11.24^{C} Print three different phase-plane graphs for the same Q value in the range 1–10, showing that the same limit cycle for van der Pol's equation appears to be approached from initial conditions both inside and outside a limit cycle in the phase plane.

EXERCISE 11.25^{P} Do the previous exercise, but for the Raleigh oscillator.

EXERCISE 11.26^{E} For two different Q values in the range 0.5–2 find the limit cycle for the van der Pol oscillator and print a phase-plane graph showing it. For each of these two cases list the Q value, the period of the limit cycle oscillation, and its amplitude $\max_t x(t)$.

EXERCISE 11.27^{E} Do the previous exercise, but for the Raleigh oscillator.

EXERCISE 11.28^{E} For some moderately large Q value such as 10, plot energy as a function of time for a limit cycle of the van der Pol oscillator on the same graph as its position $x(t)$. Interpret the result in a few expository sentences.

EXERCISE 11.29^{E} Do the previous exercise, but for the Raleigh oscillator.

EXERCISE 11.30^{E} Find significant differences between solutions of the Raleigh oscillator and those of the van der Pol oscillator for some same value of $Q \geq 1$. Describe these differences and illustrate them by graphs. Explain in qualitative physical terms how these differences arise.

EXERCISE 11.31^{E} Print phase-plane plots showing limit cycles for the van der Pol oscillator for two values of $Q \ll 1$ and similarly for the same values of Q for the Raleigh oscillator. Be careful to use sufficiently small dt to get reasonably accurate results.

EXERCISE 11.32^{H} For either the van der Pol or the Raleigh oscillator, plot the period and the amplitude of the limit cycles for $0.05 \leq Q \leq 20$ as a function of $\log Q$.

11.4 ♠ DUFF: Domains of Attraction in Nonlinear Oscillators

Initial conditions have almost no long term effects for most of the oscillators studied so far in this chapter. For almost any initial conditions (for the damped driven harmonic oscillator and for the van der Pol and Raleigh self-excited oscillators) the result after a sufficiently long time is the limit cycle motion. In this sense the entire phase plane with the exception of the origin could be called the domain of attraction of the limit cycle. The one exceptional case is the trivial solution $x(t) = 0$ which in these cases is unstable. A more serious dependence on initial conditions arises in the Duffing oscillator studied in this section.

The Duffing oscillator is (an idealized model of) a damped driven oscillator in which the spring is nonlinear and weakens when stretched too far. Linear damping is included via a viscous force $F_{\mathrm{drag}} = -b\dot{x}$ just as in the damped harmonic oscillator. But now the spring force is changed to $F_{\mathrm{spring}} = -k[x - x(x/\ell_0)^2 - Z\cos\omega t]$ so that the force weakens as x reaches an appreciable fraction of a characteristic length ℓ_0. In the absence of the driving term, this is still a conservative force with a potential energy $U = \frac{1}{2}kx^2[1 - \frac{1}{2}(x/\ell_0)^2]$. When written in terms of dimensionless variables, or units where $\omega_0^2 \equiv k/m = 1$ and $\ell_0 = 1$, the equation of motion for a Duffing oscillator becomes

$$\ddot{x} + \dot{x}/Q + x - x^3 = G\cos\Omega t \quad . \tag{11.25}$$

Since the length scale here is set by ℓ_0, the strength of the driving force $G \equiv Z/\ell_0$ is a significant dimensionless parameter, as are $Q \equiv \omega_0 m/b$ and $\Omega \equiv \omega/\omega_0$. Approximate analytic solutions of this equation are discussed, for example, in Baierlein's *Newtonian Dynamics* (mentioned on page 170).

A worksheet to solve equation 11.25 can be produced simply by editing the acceleration formulae in the work block of the **OSC** worksheet. To avoid changing a few labels you can complete the **DUFF** worksheet from the student software package instead. This worksheet can be tested by calculating solutions with $G \ll 1$ and small initial conditions. The nonlinear term should then have negligible effect on the solution, which should therefore agree with a solution for the same parameters in your previously debugged **OSC** worksheet.

Domains of Attraction

Using a moderate driving force of $G = 0.15$ explore solutions of Duffing's equation. You will find initially that it behaves much like the driven damped harmonic oscillator. Small initial conditions x_0, v_0 will evolve until a limit cycle (steady-state oscillatory motion) is reached. But very large initial conditions, e.g., $|x| \gg 1$, will lead rapidly toward $x = \infty$, showing that the domain of attraction of the limit cycle is not as large as it is in the linear case. Here **domain of attraction** means that subset of the phase plane (at $\cos\Omega t = 1$) from each point of which a solution leads arbitrarily close to the limit cycle.

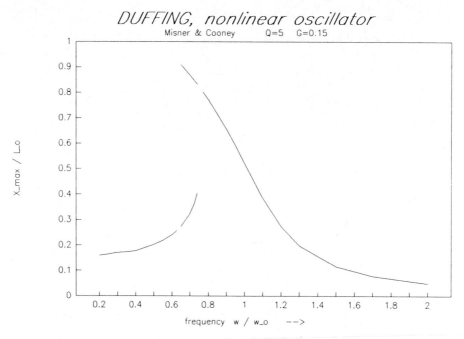

Figure 11.2: Duffing resonance curve. The resonance curve shows the amplitude of the steady-state motions produced by a driving force $F = k\ell_0 G \cos \omega t$ for the Duffing oscillator with a nonlinear spring $F_{\text{spring}} = -kx + kx^3/\ell_0^2$. The plotted quantity is the dimensionless amplitude x_{max}/ℓ_0 against $\Omega = \omega\sqrt{m/k}$. For this curve the damping is specified by $Q = 5$.

The divergence toward infinity for some initial conditions is easily understood as an artifact of our oversimplified description of a spring that weakens at large amplitude by a force law $F = -x(1 - x^2)$. This force law does weaken (compared to $F = -x$) as x grows from very small values, but for $x \geq 1$ it does quite strange things (it pushes out instead of pulling in), which the mathematics takes seriously. We have no basis in physics for the force formula at those large x values and need not assign solutions in that range any physical significance.

Careful investigation of solutions for moderate values of x does, however, reveal a surprising result. At some driving frequencies Ω there can be two distinct stable limit cycles (in addition to the degenerate cases of the unstable origin and stable infinity). Figure 11.2 is the analogue in this case of the resonance curve of Figure 11.1 (page 165) for the linear spring. For those frequencies where two possible amplitudes are shown, two stable limit cycles exist. Which one an actual oscillator mechanism chooses depends on its initial conditions. If the driving motor's frequency is slowly changed as the mechanism oscillates, the initial conditions for the new frequency are coordinates on the stable limit cycle that was reached at the previous nearby frequency. Thus the oscillator remembers whether it reached a given frequency from higher or lower frequencies when it has two possible limit cycles. This is an example of **hysteresis**.

EXERCISE 11.33^C Complete the DUFF worksheet—or build it by editing OSC—to solve the Duffing equation 11.25. Print a phase-plane graph where x remains quite small and compare it with a graph from the linear oscillator (same parameters and initial conditions) to verify that both agree for, e.g., $G = 0.0001$.

EXERCISE 11.34^H Find the potential energy $U(x)$ corresponding to a force $F = -x + x^3$ and draw a graph of U. For three different energies (negative, low positive, high positive) mark on the graph the regions of x where the kinetic energy is positive when energy $E = \frac{1}{2}\dot{x}^2 + U(x)$ is conserved ($Q \to \infty$, $G = 0$). Sketch (by hand) orbits in the phase plane corresponding to all the possible motions at these three energies.

EXERCISE 11.35^C Select from Figure 11.2 a frequency at which two different amplitudes are shown. Use your Duffing worksheet to find initial conditions that lead to each of these two different limit cycles and print phase-plane graphs of each.

EXERCISE 11.36^H Choose a driving frequency for which there are two different stable limit cycles and make a collection of sample initial data (at $\cos \Omega t = 1$) that lead to each. Record lists of such initial data on your notes page and then make a scatter plot of them (/Graph Option Format Graph Symbols), using different graph symbols for the two different classes of data according to which limit cycle they approach.

EXERCISE 11.37^E The x^3 term in the Duffing equation converts the driving frequency Ω into other frequencies (via trigonometric identities for $\cos^3 \Omega t$), most importantly the frequency 3Ω. Look therefore for something different to happen when $\Omega = 1/3$, i.e., when $3\omega = \omega_0$. Study various graphs at this frequency and see if you can identify how the motion there differs from nearby frequencies such as $\Omega = 1/4$ or $\Omega = 1/2$.

EXERCISE 11.38^H Modify the Duffing worksheet to use a nonlinear spring that stiffens at large amplitude, $F = -x(1 + x^2)$; see if there can again be two stable limit cycles at some frequencies. Plot a graph like Figure 11.2.

Chapter 12

Waves on a Hanging Rope

Most introductory physics texts discuss standing waves on a string stretched under a uniform tension T between two fixed end points. This example is universally popular for a number of reasons. First of all, it is an excellent approximation for the behavior of the strings on real musical instruments ranging from guitars to pianos. Secondly, these standing waves turn out to be mathematically identical to the standing waves that exist in the air inside a pipe, another example that is universally developed in introductory texts. In addition, the standing waves that emerge from this analysis are examples of an important class of vibrational motion, termed normal modes, common to many vibrating objects. The hallmark of such normal-mode vibration is that each point on the object undergoes simple harmonic motion with exactly the same frequency.

But what happens if the tension is not constant along the string? Does a rope or a chain exhibit standing waves when it is hanging from the ceiling under only its own weight? If so, what are their characteristic frequencies? What shape do these waves have? How do they differ from standing waves on a uniform string? Spreadsheet analysis will allow us to answer all of these questions. This hanging rope model is important because it shows, with an easily observed apparatus, phenomena that provide analogies for quantum mechanics.

12.1 The Hanging Rope

Suppose we hang a heavy rope or a chain from the ceiling so that its top end can smoothly pivot and its bottom end is free to swing around. If we shake the free end of the rope, waves travel up the rope and are reflected from the fixed upper end. Observation might suggest that these waves do not travel at a constant speed, or that they change shape or size, but such observations are difficult to make by eye. A typical interplay of theory and experiment enters now to suggest we make other observations that replace these first impressions but permit good accuracy. The theory we use is not the yet-to-be-developed theory of the new situation, but the simpler, related theory of a previously studied case—uniform tension.

You will recall that waves on a string under a uniform tension T travel at a speed $v = \sqrt{T/\mu}$ where $\mu = M/L$ is the uniform mass density, M is the total mass of the rope, and L is its length. For our hanging rope, μ is still constant along the entire length of the rope but T is not. Rather, T increases as we go from the bottom to the top. To see why this is so, we need only remember that any point on the rope must support the weight of the mass $(h/L)M$ of the rope below it. Thus the tension in the rope a distance h above its bottom end is

$$T = (h/L)Mg = \mu gh. \qquad (12.1)$$

This suggests that the wave speed $v = \sqrt{T/\mu} = \sqrt{gh}$ increases because the tension increases as we move up the rope. But this is still hard to check by simple observation.

Measuring Wave Velocity Indirectly

Wave speed v in the constant tension case could be found indirectly from $v = \lambda \nu$ by measuring the wavelength λ and frequency ν of a standing wave or normal mode. The suggested experiment then is to try to set up a normal mode in the hanging rope where every part moves with the same frequency. If the wave velocity increases with height, then so will the wavelength from $v = \sqrt{T/\mu} = \sqrt{gh} = \lambda \nu$ or $\lambda = \sqrt{gh}/\nu$. Such a variation in wavelength in a standing wave is easy to check qualitatively and is not difficult to measure.

Some idealizations we have implicitly made here are harder to realize in the hanging rope than in the string under external tension, however. The uniform tension formulae that we use assume that forces on a piece of rope or string arise only from the varying directions of the force due to tension at different points on the string. When tension is high, as on a violin string, this is a good approximation. The principal neglected forces are the stiffness of the string (does it require force to bend it?) and air resistance when it moves. For a light cord hanging under its own weight the tension may not be high enough to dominate these other forces. Thus a high density flexible chain is better. Two meters of solid gold jewelry chain might be ideal, but we haven't had the chance to try that. The type of beaded chain used to keep an old-fashioned bathtub stopper from getting lost works quite well. Use a small variable speed motor on a ring stand clamp to rotate the top end of the light chain in a small circle (a centimeter or less) and vary the frequency to search for steady patterns (normal modes). See how well the $\lambda \propto \sqrt{h}$ relationship works, if at all, and then proceed with the more detailed theory that follows.

Recalling that $T = \mu g h$ (for small amplitude waves), we can further simplify this to give our final differential equation for $R(h)$:

$$\frac{d}{dh}\left[h\frac{dR}{dh}\right] = -(\omega^2/g)R(h) \quad . \tag{12.9}$$

Now all we have to do is build a worksheet to solve this equation. This is done by rewriting this equation as two first-order equations; we make a definition $p = h(dR/dh) = T_r/\mu g$ and then update p and R on each row of the worksheet according to

$$\begin{aligned} dR &= p\,dh/h \\ dp &= -(\omega^2/g)R(h)\,dh \quad . \end{aligned} \tag{12.10}$$

EXERCISE 12.1[C]   Repeat the arguments of this section and find replacements for equations 12.9 and 12.10 in the simpler case where gravity can be neglected, and a constant tension T is maintained on the rope by pulling on its ends (e.g., a jump rope). Use dimensional analysis (what constants appear essentially in your equations) to define a base frequency ω_0 in this case for a rope of length L.

EXERCISE 12.2[C]   Use the free body diagram for the segment of rope shown in the right hand part of Figure 12.1 to calculate the rectangular components of the net force acting on this segment of rope. Sketch diagrams that resolve each force into its horizontal and vertical components using the angles defined in the diagram. Derive equations 12.4–12.7 using the approximations from footnote 2 on page 180.

12.3   HROPE: **Finding the Normal Modes**   .

From the problem of the uniform string recall that standing wave solutions only exist for particular values of the frequency. The values of these allowed frequencies were determined by the boundary conditions which required that each end of the stretched string be fixed in space, making each end a node of the vibrational motion.

Eigenvalues

The boundary conditions for the hanging rope restrict the values of ω in a similar way. In addition to satisfying equations 12.9 or 12.10, our standing wave solutions must also conform to the requirements that the upper end of the rope is fixed, so $R(L) = 0$, and that the tension at the lower end is zero since this end is free of all external forces. When we construct our spreadsheet to solve equation 12.9 we have to be sure that each solution we obtain meets these two criteria. Only proper values of ω will produce physically proper solutions that satisfy these particular boundary conditions. Such proper values and proper solutions are sometimes called the **eigenvalues** and **eigenfunctions** of the differential equation consistent with a particular set of boundary conditions.

labels:	h	R	phs	a
init	h_0:	R_0:	$p_{1/2}$:	a_0:
data:	L	0	$h_0 s_0 + a_0\, dh/2$	$-(\omega^2/g)R_0$
typical	h_1:	R_1:	$p_{3/2}$:	a_1:
line:	$h_0 + dh$	$R_0 + \dfrac{2p_{1/2}\, dh}{h_0 + h_1}$	$p_{1/2} + a_1\, dh$	$-(\omega^2/g)R_1$

Schema 12.1: Procedure to solve the variable tension eigenvalue equation.

Differential equations subject to boundary conditions like these often have valid solutions only for particular values of certain physical parameters. This fact lies at the heart of the discrete character of many quantities in quantum mechanics. For example, an electron in an atom is described by a quantum mechanical wave. It is confined to the neighborhood of the nucleus by the way its wavelength (controlled by, not tension as here, but by electrostatic potential) varies with distance from the nucleus. This wave also vibrates in normal modes whose frequencies in certain combinations give the frequencies of light that the atom can emit or absorb. These frequencies are the main direct observational data that we have about atoms.

Numerical Approximation

To build a worksheet to solve equation 12.9, we use the quantity $p = h\, dR/dh$, which can be called the pull on the rope. (Physically p is just the horizontal component of the tension, $T_r = \mu g h\, dR/dh$ but without the constant factors μg.) This allows us to rewrite equation 12.9 as $dp/dh = a$, where $a = -(\omega^2/g)R(h)$ is a dimensionless centripetal acceleration. We can then use the leapfrog integration scheme to solve equations 12.10 down the length of the rope. The details of this numerical solution are given in Schema 12.1.

Input and Output

The quantities we need to provide on the I/O page of the worksheet are the length of the rope L, the number of steps n into which we will divide the rope to solve our differential equation, the acceleration g due to gravity pulling down on the rope, the amount $(dR/dh)_L$ that the top of the rope slopes away from the vertical, and the angular frequency ω of the rope's circular motion. While the significance of L, n, and g are obvious, the role of $(dR/dh)_L$ and of ω merit a bit more consideration.

The angular displacement of the top of the rope, $(dR/dh)_L$, determines the overall amplitude of the standing wave on the rope. In an experimental study of the standing waves on a real rope, we would want to keep this amplitude small to match the assumptions of our model. For calculations here any nonzero value of $(dR/dh)_L$ will do since the small-amplitude condition is assumed in

```
D3: PR 'HROPE, Waves on a Hanging Rope                          READY
```

	A	B	C	D	E	F	G	H
1	your name & today's date
2	your class & section
3				HROPE, Waves on a Hanging Rope
4	Description:
5			Try various values of (w/w_p); normal modes satisfy the
6			boundary condition at the bottom end (height = 0) of the rope
7			that the restoring "pull" (= h dR/dh) should vanish there.
8	---
9	Construction parameters:
10		Length:		2 meter		PendFreq:	0.352 Hz
11		Gravity:		9.8 m/s^2
12	Input Boundary Conditions:
13		OscFreq/PendFreq:		7.450 (=w/w_p)
14		OscFreq:		2.625 Hz		w:	16.5 rad/sec
15		Top displacement:		0 m
16		Top slope dR/dh:		0.1 radian
17	Approximation:
18		Number of steps:		200		dh:	-0.01 m
19	Results:
20		Bottom "pull":		0.0000 m		Number of nodes:		5

Screen 12.1: I/O page of the HROPE worksheet.

our differential equation and the solution is simply proportional to $(dR/dh)_L$.

The angular frequency of the rope's circular motion plays a much more interesting role in our analysis of these standing waves. It will prove convenient to express ω in terms of some natural frequency for our system. Since the solution to our differential equation does not depend on μ and is linear in $(dR/dh)_L$, ω can only depend on L and g. Dimensional analysis then tells us that ω must be proportional to $\sqrt{g/L}$. Thus, if we express ω in the natural unit of angular frequency, $\omega_p = \sqrt{g/L}$, the values we obtain for the ratio (ω/ω_p) corresponding to physically acceptable solutions of equation 12.9 will be valid for any value of L or g.

Screen 12.1 shows the I/O page for HROPE, the hanging rope worksheet. It contains reasonable starting values for the input parameters and displays a selection of derived quantities that characterize our system. The first few lines of the work block are shown in Screen 12.2, using the formulas in Schema 12.1. The typical-line formulas have been copied down 199 rows to yield values of h ranging from L (at the top) to 0 (at the bottom of the rope). Figure 12.2 shows the shape of the standing wave on the rope for the parameter values given in Screen 12.1.

Searching for Normal Modes

How are we to determine if the solution we have just obtained satisfies our boundary conditions? The requirement $R(L) = 0$ is built into our worksheet and so is automatically satisfied by any solution that our worksheet generates.

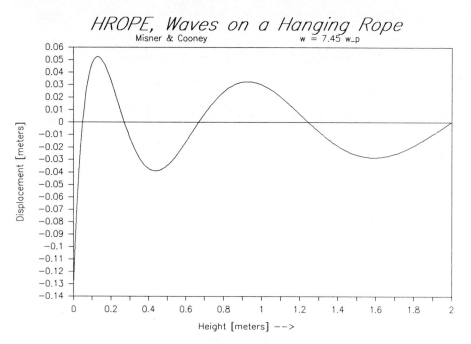

Figure 12.2: A standing wave on the hanging rope.

However, the boundary condition for the bottom of the rope is generally *not* satisfied by solutions that the worksheet produces. Rather, it is up to us to vary (ω/ω_p) and check if the pull on the end of the rope turns out to be zero. Only values of (ω/ω_p) that produce $p(0) = 0$ are physically acceptable solutions.

To observe the behavior of p as we vary (ω/ω_p), it is necessary to have values of p_0, p_1, p_2, ... rather than the values of $p_{1/2}$, $p_{3/2}$, $p_{5/2}$, ... tabulated as phs in our worksheet. To accomplish this, we have added a fifth column to HROPE; it calculates $p_1 = (p_{1/2} + p_{3/2})/2 = p(L - |dh|)$ and so on down to $p_{200} = p(0)$ with $p_0 = p(L)$ as its first entry. This column is labeled p in Screen 12.2. The value of p_{200} is also displayed on our I/O page (Screen 12.1) to allow us to see this crucial boundary condition value as we vary (ω/ω_p). The particular values shown in these two screens correspond to a valid standing wave solution having five nodes.

There are, in fact, many values of (ω/ω_p) that satisfy our boundary conditions. The smallest of these eigenvalues corresponds to the rope's swinging back and forth like a long, floppy pendulum. This mode has only one node, located at the top of the rope. As we continue to increase (ω/ω_p), we find the next eigenvalue for which $p(0)$ is again zero. This eigenvalue corresponds to a standing wave that has two nodes, one at the top and another near (but not at) the bottom of the rope. As we continue to increase (ω/ω_p) we find a whole series of these eigenvalues, each corresponding to a standing wave eigenfunction having one more node than its predecessor. Graphs of R versus h display,

```
X174: PR +TOP_SLOPE*H                        EDIT
+D16*T174
```

	T	U	V	W	X
171	height	displ	pull	-w^2 R/g	Plot_pull
172	[m]	[m]	[m]	[1]	[m]
173	h	R	phs	a	p
174	2	0	0.2	0	0.2
175	1.9900	−0.00100	0.199721	0.027820	0.199860
176	1.9800	−0.00200	0.199164	0.055742	0.199443
177	1.9700	−0.00301	0.198327	0.083727	0.198745
178	1.9600	−0.00402	0.197209	0.111737	0.197768
.					
.					
.					
372	0.0200	−0.06603	0.045193	1.832485	0.054355
373	0.0100	−0.09616	0.018507	2.668603	0.031850
374	−0.0000	−0.13317	−0.01845	3.695818	0.000028
375					

Screen 12.2: The work block of HROPE.

in turn, the shape of each of the eigenfunctions.

EXERCISE 12.3[C]   Using the HROPE worksheet provided (or one that you have created) find the values of (ω/ω_p) for the four lowest frequency normal modes of the hanging rope. For each of these modes, print a graph showing the shape of the standing wave. Label each of these graphs with its eigenvalue (ω/ω_p).

EXERCISE 12.4[C]   Describe how the shape of these four standing waves differ from the shape of the first four standing waves on a uniform stretched string. Graphically compare the corresponding eigenfunctions.

EXERCISE 12.5[P]   For each standing wave you have identified, plot the pull and the displacement on the same graph. Comment on the relation between these pairs of data.

EXERCISE 12.6[P]   Characterize the motion of the free end of the rope. Do your results suggest why crane halyards always carry a substantial weight on their lower end?

EXERCISE 12.7[E]   Tabulate and graph the pull on the bottom of the rope for values of (ω/ω_p) ranging from 7.5 to 15.0. The zero-crossings of this graph will help you "zero in" on the next few higher normal modes of the rope.

EXERCISE 12.8[E]   Make phase-plane plots for this system by plotting p versus R. Explore how these graphs change as you vary (ω/ω_p). Describe how you could distinguish allowed solutions from unphysical solutions by an examination of such phase-plane plots.

EXERCISE 12.9[H]   How sensitive are your eigenvalues for (ω/ω_p) to the number of integration steps you use in solving the differential equation describing the motion of the rope?

EXERCISE 12.10^H Find, using the library and more advanced texts, a solution to this hanging rope problem that solves equation 12.9 using Bessel functions. (You need not present or master that solution.) Compare your results for the eigenvalues of (ω/ω_p) with those obtained analytically.

EXERCISE 12.11^E Find numerical solutions of the equations you derived in Exercise 12.1, assuming that the ends are held fixed. Do this by modifying the HROPE worksheet. (If you want to save your modifications in the same compacted form in which HROPE is distributed, see Appendix C, Section C.1 for the use of the {Alt-S} macro, which does this job.) Find the four lowest eigenvalues (frequencies) expressed as multiples of the base frequency ω_0 you defined in Exercise 12.1 and print out a graph showing the eigenfunction $R(h)$ with four nodes (including the two ends).

EXERCISE 12.12^E Derive the same results as those in the previous problem analytically, i.e., express the eigenfunctions $R(h)$ as familiar functions by solving the replacement for equation 12.9, which (except for variables with different names) you have met before, and give a simple formula for the nth eigenvalue ω_n/ω_0.

12.4 Listings

```
      TITLE1= D3: 'HROPE, Waves on a Hanging Rope
          L=D10: U 2                        W_P=G10: (F3) @SQRT(G/L)/2/@PI
          G=D11: U 9.8
    |W/W_P|=D13: (F3) U 7.45
       FREQ=D14: (F3) +|W/W_P|*W_P        W=G14: (F1) 2*@PI*FREQ
  TOP_DISPL=D15: 0
  TOP_SLOPE=D16: U 0.1
          N=D18: 200                      DH=G18: -L/N
            D20: (F4) +END_PULL              H20: +NUM_NODES

NOTES_BLOCK=I21..R90
     TITLE2=J21: +A1&"     w = "&@STRING(|W/W_P|,2)&" w_p"
   |W^2/G|=K22: (F2) +W*W/G

  NUM_NODES=L28: @DCOUNT(DATA,0,CRITERION)
                   DATA=T173..X374
               CRITERION=K25..K26
           K25: ^R
           K26: (T) +U173*R<0#OR#R=0

#=T171
        H=T174: +L                    T175: (F4) +H+$DH
        R=U174: +TOP_DISPL            U175: +R+2*PHS*$DH/(H+T175)
      PHS=V174: +TOP_SLOPE*H+A*DH/2    V175: +PHS+W175*$DH
        A=W174: -$|W^2/G|*R            W175: -$|W^2/G|*U175
        P=X174: +TOP_SLOPE*H          X175: (PHS+V175)/2
                            END_PULL=X374: (V373+V374)/2
```

Listing 12.1: Important parts of the HROPE worksheet. The function
@DCOUNT used in cell L28 to scan the work block and count the number of
nodes in $R(h)$ is an advanced database statistical function. If you wish
to learn how to use it you must consult another spreadsheet reference or
use {Help}. It counts the number of rows in the range DATA that satisfy
the conditions specified in CRITERION. Our criterion is that the value in
the R column is either zero or has changed sign from the row above.

Chapter 13

Lorentz Force

The electromagnetic $(\mathbf{E}, \mathbf{B})$ field describes one of the four well established forces that play a fundamental role in the structure of matter and the universe. It is the only fundamental force apart from gravity that operates on the laboratory scale. The other two fundamental forces (the weak and the strong nuclear forces) act only on subatomic scales. The electromagnetic force $\mathbf{F}$ acting on a particle of charge q is given by the Lorentz force law

$$\mathbf{F} = q(\mathbf{E} + \mathbf{v} \times \mathbf{B}) \tag{13.1}$$

where $\mathbf{v}$ is the velocity of the charged particle. Thus $\mathbf{E}$ describes the velocity-independent part of the force, and $\mathbf{B}$ the velocity-dependent part. The worksheets described in this chapter solve for particle motions allowed by the force law 13.1 when the $\mathbf{E}$ and $\mathbf{B}$ fields are specified. Maxwell's electromagnetic field equations restrict the possible configurations of $\mathbf{E}$ and $\mathbf{B}$ and show what is needed to produce these fields. We shall verify them as needed.

In numerical approximations to the motions allowed by equation 13.1 or

$$m\frac{d\mathbf{v}}{dt} = q(\mathbf{E} + \mathbf{v} \times \mathbf{B}) \tag{13.2}$$

we will, as usual, want to monitor our approximations by looking at energy conservation. In this connection it is important to know that magnetic forces do no work:

$$W_B = \int q\mathbf{v} \times \mathbf{B} \cdot d\mathbf{r} = 0 \tag{13.3}$$

since $\mathbf{v} \times \mathbf{B}$ is perpendicular to the displacements $d\mathbf{r} = \mathbf{v}\,dt$ that occur during the motion. The work done by the electric force $q\mathbf{E}$ can be accounted as accompanied by a decrease in the electrostatic potential energy $U = qV$ where the electrical potential V satisfies $dV = -\mathbf{E} \cdot d\mathbf{r}$. As a consequence the total energy

$$\mathcal{E} = \tfrac{1}{2}m\mathbf{v} \cdot \mathbf{v} + qV \tag{13.4}$$

remains constant when equation 13.2 is solved exactly.

EXERCISE 13.1^{E} By differentiating equation 13.4 and using equation 13.2, prove the conservation of energy law for electromagnetic forces.

13.1 ELMAG: Charged Particle in a Magnetic Field

Although the motion of a charged particle in a uniform magnetic field **B** can easily be solved analytically, it is appropriate to begin with that problem in our numerical explorations to be sure that our numerical methods are working and to become acquainted with the relevant dimensional analysis so that computations can be scaled efficiently. We will then add a constant electric field and let numerical experiments suggest for the motion in that case an expression that can be checked analytically.

Dimensional Analysis

When **E** $= 0$, equation 13.2 can be rewritten as

$$\frac{d\mathbf{v}}{dt} = \left(-\frac{q\mathbf{B}}{m}\right) \times \mathbf{v} \tag{13.5}$$

which can be interpreted as saying that the velocity vector **v** rotates with angular velocity $\boldsymbol{\omega} = -q\mathbf{B}/m$. When the vector **B** is constant (independent of space and time) the consequences are easy to describe. The velocity component $v_\parallel$ parallel to **B** is constant, and the rotating velocity component $v_\perp$ orthogonal to **B** corresponds to circular motion in the plane perpendicular to **B**. The result is a helical or spiral motion of the charged particle about a magnetic field line. Let **B** define the x-axis. Then in the yz-plane there will be a circular motion with a centripetal acceleration $v_\perp^2/r$ so that equation 13.5 reads $v_\perp^2/r = (qB/m)v_\perp$. When expressed in terms of the angular velocity $\omega = v_\perp/r$ of the circular motion this becomes $\omega = qB/m$, so the circular motion equations are satisfied, and the angular velocity is even independent of r. In the work block of a numerical solution we therefore take this angular frequency

$$\omega_c \equiv qB/m \tag{13.6}$$

as the unit for angular velocity or, equivalently, take m/qB as our unit of time.

The radius of the circular motion in the plane perpendicular to **B** is given by $\omega = v_\perp/r$ or $r = v_\perp/\omega_c = mv_\perp/qB$. The natural unit of length therefore depends on the momentum $mv_\perp$ of the particle. In the worksheet we use to reproduce this motion we therefore take the particle's mass m as a mass unit, the intended typical initial velocity C as setting the velocity and momentum units, and thus the length unit becomes $r_c \equiv mC/qB$.

When an electric field is included we need a suitable unit for it. From the Lorentz force law $\mathbf{F} = q(\mathbf{E} + \mathbf{v} \times \mathbf{B})$ it is clear that the dimensions of an electric field are those of vB; we therefore choose to measure E in units of CB where C is our unit of velocity.

Numerical Methods

Because the force law 13.1 is velocity dependent, we cannot use the simple leapfrog method based on the **LEAP** worksheet that has served us well in many applications. Instead we adapt the **RK2** worksheet from Section 6.4. The student software package has a copy of this (called **ELMAG.wk1**) in which the column headings and graph labels are already prepared. But you will still have to copy and edit the columns that integrate a one-dimensional equation $md^2y/dt^2 = F(y, v, t)$ to produce a second set of computations for the z-coordinate. (The magnetic field direction, taken as the x-axis, can be omitted since the motion there is just constant velocity.) The interesting electric field to include is one perpendicular to the magnetic field. Thus one sets $B_x = B$ with $B_y = 0 = B_z$ and $E_z = E$ with $E_x = 0 = E_y$ in equation 13.2 to find the acceleration

$$
\begin{aligned}
a_z &= (q/m)(E_z - v_y B_x) \\
&= (qB/m)(E/B - v_y) \\
&= (qB/m)[(E/CB)C - v_y]
\end{aligned}
\tag{13.7}
$$

which, in the units we suggest $(qB/m = C = CB = 1)$, can be programmed as simply

$$
\begin{aligned}
a_y &= +v_z \\
a_z &= E - v_y
\end{aligned}
\tag{13.8}
$$

Exploration

After completing the **ELMAG** worksheet, set the electric field to zero at first and study the accuracy obtainable in reproducing the circular motion expected from the analytic solution. Look at the constancy of the energy for large time steps $\omega_c dt \approx 1$ as well as smaller ones. Then, choosing a time step small enough to give moderate accuracy, look at the effects of a transverse electric field. Try small values of $E \sim 0.02\,CB$ as well as larger ones, $E \sim 1\,CB$. Then attempt to describe analytically what you see. If $\mathbf{r}_c(t)$ and $\mathbf{v}_c(t)$ describe the circular motion in the yz-plane that occurs when $E = 0$, what is the difference between these circular motions and the total motion $\mathbf{r}(t)$ and $\mathbf{v}(t)$ that is plotted when you calculate the worksheet with a nonzero E?

EXERCISE 13.2$^\text{C}$ Complete the **ELMAG.wk1** worksheet and test it. Then print out a listing of the equations you have used in the first two rows of computation in the work block. If you want to save your modifications in the same compacted form in which **ELMAG** is distributed, see Appendix C, Section C.1 for the use of the {Alt-S} macro, which does this job.

EXERCISE 13.3^C Explore the effect of changing the step size dt in the ELMAG worksheet you constructed in the previous exercise, using $E = 0$. Print two graphs: one should show the orbits as closed circles by using an adequately small value of dt; the other should show moderate energy errors when a larger dt is chosen. Be sure to write on each graph (or include in its titles) the value of the step size $\omega_c\, dt$ that is used to produce it.

EXERCISE 13.4^C Continuing with the same worksheet, print two orbit graphs (using adequately small dt) showing the effect of an electric field **E**. One graph should be for a small value of E, one for a large value. Be sure the values of E/CB used are marked on each graph. Can you find a particularly simple motion when $E = CB$ and $v_\perp = C$?

EXERCISE 13.5^E Try to describe the effects of a transverse electric field in equations. Write $\mathbf{v}(t) = \mathbf{v}_d(t) + \mathbf{v}_c(t)$ where $\mathbf{v}_c(t)$ satisfies equation 13.5 and $\mathbf{v}_d$ represents the additional consequences of the electric field that you saw in the preceding exercise. Try to write an equation or mathematical statement that specifies $\mathbf{v}_d(t)$. How is the direction of $\mathbf{v}_d$ related to **E** and **B**?

EXERCISE 13.6^H Prove analytically that, with the formula for $\mathbf{v}_d$ you wrote in the previous exercise, an exact solution of equation 13.2 is $\mathbf{v}(t) = \mathbf{v}_d(t) + \mathbf{v}_c(t)$.

13.2   ♠ EM2: Magnetic Leapfrog Integration

We would like to be able to solve numerically for the motion of charged particles in nonuniform magnetic fields, such as the earth's dipole field, but the accuracy obtained using the Runge-Kutta (RK2) method in the ELMAG worksheet was discouraging. With that method the energy increased exponentially, so the results were qualitatively misleading unless quite small step sizes were used yielding only a few cyclotron orbits (circles around the magnetic field line). In this section we develop a leapfrog method that can be used for magnetic forces and find that it does conserve energy in the case of uniform magnetic fields. It is then applied in the next section to a simple but illustrative nonuniform magnetic field.

If equations 13.8 are rewritten as difference equations, which calculate how the velocity changes from one worksheet row to the next, they read

$$
\begin{aligned}
v_{y,n+1/2} &= v_{y,n-1/2} + (v_z B)_n\, dt \\
v_{z,n+1/2} &= v_{z,n-1/2} + (E - v_y B)_n\, dt \quad .
\end{aligned}
\tag{13.9}
$$

In the uniform field case we can set $B = 1$ but it is kept B here to remind us that in a generalization we will need a function of position $B_x(x, y, z)$ evaluated at the current position (x_n, y_n, z_y) of the particle. The difficulty with these equations that prevents our using the leapfrog method from Section 6.4 is the appearance of v_n on the right hand sides. We attack that difficulty here by writing, for each of the y- and z-components of **v**,

$$
v_n = \tfrac{1}{2}(v_{n+1/2} + v_{n-1/2}) \quad .
\tag{13.10}
$$

The challenge now is to solve equations 13.9 for the new velocities, which appear on both sides of the equations, in terms of the old. First move all the new velocity terms to the left hand sides:

$$v_{y,n+1/2} - v_{z,n+1/2}B\,dt/2 = v_{y,n-1/2} + v_{z,n-1/2}B\,dt/2$$

$$v_{z,n+1/2} + v_{y,n+1/2}B\,dt/2 = v_{z,n-1/2} - v_{y,n-1/2}B\,dt/2 + E\,dt \quad.$$

$$(13.11)$$

This is best solved by writing it first as a matrix equation:

$$\begin{pmatrix} 1 & -\frac{1}{2}B\,dt \\ \frac{1}{2}B\,dt & 1 \end{pmatrix} \begin{pmatrix} v_y \\ v_z \end{pmatrix}_{n+\frac{1}{2}} =$$

$$= \begin{pmatrix} 0 \\ E \end{pmatrix}_n dt + \begin{pmatrix} 1 & +\frac{1}{2}B\,dt \\ -\frac{1}{2}B\,dt & 1 \end{pmatrix} \begin{pmatrix} v_y \\ v_z \end{pmatrix}_{n-\frac{1}{2}} \quad.$$

$$(13.12)$$

This equation can be solved by multiplying by the inverse of the matrix that appears on the left. This inverse is

$$\begin{pmatrix} 1 & -\frac{1}{2}B\,dt \\ \frac{1}{2}B\,dt & 1 \end{pmatrix}^{-1} = \frac{1}{1+(B\,dt/2)^2} \begin{pmatrix} 1 & +\frac{1}{2}B\,dt \\ -\frac{1}{2}B\,dt & 1 \end{pmatrix} \quad (13.13)$$

which leads finally to the desired solution

$$v_{x,n+1/2} = v_{x,n-1/2}$$

$$v_{y,n+1/2} = \left[v_{y,n-1/2}D' + v_{z,n-1/2}B\,dt + \tfrac{1}{2}EB\,dt^2 \right]/D \qquad (13.14)$$

$$v_{z,n+1/2} = \left[v_{z,n-1/2}D' + (E - v_{y,n-1/2}B)dt \right]/D \quad.$$

where $D \equiv 1+(B\,dt/2)^2$ is the determinant that arises when we invert a matrix and $D' \equiv 1 - (B\,dt/2)^2$ is a similar quantity that arises in squaring a matrix. In these equations we have also included the statement that $v_x = $ const, which we know from the vanishing of the x-component of the Lorentz force in this case.

When we build a worksheet to solve for Lorentz force motions using equations 13.14, it is reasonable to try using the approximation

$$\mathcal{E} = \tfrac{1}{2}\mathbf{v}_{n+1/2} \cdot \mathbf{v}_{n-1/2} - Ex \qquad (13.15)$$

for the energy, since this form was found (in Exercise 6.5 on page 81) to be exactly conserved in the absence of magnetic fields but with a constant force like $q\mathbf{E}$ here.

EXERCISE 13.7^C A worksheet template EM2.wk1 is available in the student software package that has labels and graphs designed for solving the difference equation system 13.14. Complete the worksheet, test it, and print out a listing of the formulae you programmed into the first two rows of computations in the work block. If you want to save your modifications in the same compacted form in which EM2 is distributed, see Appendix C, Section C.1 for the use of the {Alt-S} macro that does this job.

EXERCISE 13.8^C Use the worksheet from the previous problem to explore, in the simpler case $E = 0$, how the energy and period of the cyclotron motion depend on the time step $\omega_c\, dt$. Present your results as tables printed from the notes page where you collect relevant data or as graphs from these tables.

EXERCISE 13.9^E Repeat the previous exercise but with $E \neq 0$. Because of the $\mathbf{v}_d = (\mathbf{E}\times\mathbf{B})/B^2$ drift, the orbit is not periodic. The z motion, however, does remain periodic and lets you measure the period of the cyclotron motion.

EXERCISE 13.10^E In the previous exercise you should have found that the computed energy was only constant over the long term, but not over the short term. This suggests that it differs by some short term oscillation from a constant. Since the energy expression 13.15 is constant when $E = 0$ and also (from Exercise 6.5) when $B = 0$, any such oscillating adjustment to the energy formula must be proportional to both E and B. The only scalar we can form from E and B (since $\mathbf{E}\cdot\mathbf{B} = 0$) is $\mathbf{v}\cdot\mathbf{E}\times\mathbf{B}$. Therefore try adding a term $k\, dt^n \mathbf{v}\cdot\mathbf{E}\times\mathbf{B}$ to equation 13.15 and, by numerical experiments, find values for k and n that will make the resulting approximate energy formula give an exactly conserved energy for the integration procedure used in the EM2 worksheet.

13.3 ♠ EMQD: Quadrupole Magnet

Now that an efficient numerical method has been prepared in the preceding section, we can look at the motion of charged particles in nonuniform magnetic fields. There are many important applications. The electron beam that paints the picture on the face of most TV tubes is directed by magnetic fields. The van Allen belts above the earth consist of charged particles trapped in the earth's magnetic field. When they are disturbed by magnetic weather changes blown out from the sun, they can dip into the stratosphere and produce the aurora borealis. The magnetic "bottles" used in research on thermonuclear fusion power generation provide further examples. But even the simplest of these magnetic fields is rather complicated to describe mathematically—the earth's dipole field (its smoothest long range approximation) is proportional to the earth's magnetic moment μ_E and is given by

$$
\begin{aligned}
B_x &= 3\mu_E zx/r^5 \\
B_y &= 3\mu_E zy/r^5 \\
B_z &= (\mu_E/r^5)(3z^2 - r^2)
\end{aligned}
\tag{13.16}
$$

which would lead to long computations in a worksheet. To show qualitatively the effects of nonuniform magnetic fields on charged particle motions, we choose a slightly artificial example that has the advantage of being two dimensional. We will analyze the effects of a magnet whose field lines lie entirely in the xy-plane and which are the same at every value of z. The formulae for $\mathbf{B}$ then are:

$$
B_x = -ky \quad , \quad B_y = -kx \quad , \quad B_z = 0 \quad .
\tag{13.17}
$$

This field is plotted in Figure 13.1. We want to see to what extent a particle

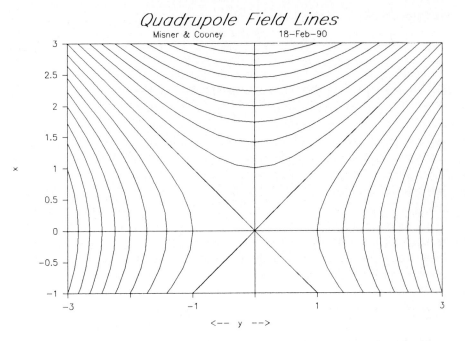

Figure 13.1: Magnetic quadrupole field. The field lines drawn here in the xy-plane show a field **B** that is zero at the origin and that can be produced by long wires outside the region shown. Two wires parallel to the z-axis carrying current directed into the page are needed above and below the center of the diagram; another two wires carrying current directed out of the page are needed to the right and left of the center.

can move in small circles around one of these field lines while sliding along it in a motion suggested by our findings for the case of constant **B**. This requires generalizing equations 13.14 to allow **B** to point in directions other than along the x-axis. We do this after we digress to a discussion of Maxwell's equations, which you may skip if your text has not yet reached that subject.

Maxwell's Equations

Maxwell's equations show what currents are required to produce the magnetic fields we want and are therefore basic engineering principles for the design of electromagnets. They also restrict the magnetic fields that are possible to realize in a vacuum (e.g., between the pole faces of an electromagnet). We want to check that equations 13.17 are a possible magnetic field configuration. The two equations we need to check are Ampere's law

$$\oint \mathbf{B} \cdot d\mathbf{r} = \mu_0 I \qquad (13.18)$$

and the magnetic Gauss's law

$$\oint \mathbf{B} \cdot d\mathbf{A} = 0 \quad . \tag{13.19}$$

In equation 13.18, I is the current encircled by the closed loop over which the integral is taken, and μ_0 is a constant that defines the Ampere (unit of current) in the SI system of units. We need only the case $I = 0$. To verify equation 13.18 for the quadrupole field of equations 13.17, we first calculate the integrand $\mathbf{B} \cdot d\mathbf{r} = -k(y\,dx + x\,dy) = -d(kxy)$. Since the integrand is a total differential (of the function $-kxy$), the value of the integral will be the difference in the values of this function at the two ends of the range of integration. But on a closed loop the two ends are the same point, so the difference is zero, and we have verified Ampere's law. Gauss's law requires a bit more calculation. Consider as a sufficiently general closed surface a rectangular box with one corner at the origin $(x, y, z) = (0, 0, 0)$ and the other corner at a point (a, b, c). On the two end faces with $z = 0, c$ the outward pointing area element $d\mathbf{A}$ has only a z-component, $dA_z = \pm dx\,dy$, but $B_z = 0$, so $\mathbf{B} \cdot d\mathbf{A} = 0$ there, i.e., the lines of $\mathbf{B}$ do not cross the $z = $ const planes. Next consider the $x = 0$ and $x = a$ sides of the box. There the nonvanishing component of the area vector is $dA_x = \pm dy\,dz$ so the flux integral over one of these sides is $\pm \iint B_x\,dy\,dz = \pm c \int_0^b (-ky)\,dy = \mp ckb^2/2$. Since this is the same, apart from the sign of dA_x, for both values of x, the two sides cancel. (A flux goes in one side and out the other.) An identical argument holds for the two remaining sides of the box, so the net flux out of the box is zero as the Gauss law demands.

Difference Equations

We now want to write equations 13.14 in a way that will allow the magnetic field to have more than an x-component. One way to do this is to recast equations 13.14 into vector notation. They can then be verified by selecting the x-axis to coincide with the direction of the $\mathbf{B}$ field at the point we select for checking but will have the same content in any rectangular axes we want to use. For example, the determinant that arose when we inverted a matrix earlier was $D = 1 + (B_x\,dt/2)^2$ in the case where $\mathbf{B}$ was entirely in the x-direction. But this must be a scalar quantity and, when it is written $D = 1 + \mathbf{B}^2 dt^2/4$ with $\mathbf{B}^2 \equiv \mathbf{B} \cdot \mathbf{B}$, it can be evaluated without needing to choose the x-axis to agree with the direction of $\mathbf{B}$. In this way we find a general expression for a magnetic leapfrog difference equation. To write it more easily we use $\mathbf{v}$ to mean $\mathbf{v}_{n+1/2}$ and $\mathbf{u}$ to mean $\mathbf{v}_{n-1/2}$. It reads:

$$\begin{aligned}
\mathbf{v} = \ & D^{-1}\big\{\mathbf{u}D' + dt(\mathbf{E} + \mathbf{u} \times \mathbf{B}) \\
& + \tfrac{1}{2}dt^2\left[\mathbf{E} \times \mathbf{B} + \mathbf{B}(\mathbf{B} \cdot \mathbf{u}) + \tfrac{1}{2}dt\,\mathbf{B}(\mathbf{E} \cdot \mathbf{B})\right]\big\}
\end{aligned} \tag{13.20}$$

where

$$D = 1 + \mathbf{B}^2 dt^2/4 \quad , \quad D' = 1 - \mathbf{B}^2 dt^2/4 \quad . \tag{13.21}$$

Some of the dt^2 terms here need explanation. The $\frac{1}{2}dt^2\mathbf{E}\times\mathbf{B}$ term, for instance, is just the term $\frac{1}{2}EB\,dt^2$ that appeared in the v_y equation in set 13.14. But there $\mathbf{E}$ was in the z-direction and $\mathbf{B}$ was in the x-direction, so the $\frac{1}{2}dt^2\mathbf{E}\times\mathbf{B}$ expression puts exactly the right quantity into exactly the right equation to agree with equations 13.14. The $\frac{1}{2}dt^2\mathbf{B}(\mathbf{B}\cdot\mathbf{u})$ term will influence only the v_x equation when $\mathbf{B}$ is in the x-direction; it converts the D' in the numerator into a D to cancel the denominator in equation 13.20 and thus agrees with the v_x equation from set 13.14. The last term, $(1/4)dt^3\mathbf{B}(\mathbf{E}\cdot\mathbf{B})$, doesn't appear in equations 13.14 since $\mathbf{E}\cdot\mathbf{B}=0$ there. We find it (for completeness sake, since it also doesn't appear in the quadrupole magnet problem) by repeating the calculations of the previous section with a nonzero E_x component, which makes the v_x equation from set 13.14 read $v_{x,n+1/2}=v_{x,n-1/2}+E_x\,dt$.

Dimensional Analysis

After we specify the quadrupole magnetic field of equations 13.17 and a constant electric field E in the z-direction, equation 13.2 includes

$$
\begin{aligned}
a_z &= (q/m)(E-v_xkx+v_yky) \\
&= (qk/m)(E/k-v_xx+v_yy) \quad .
\end{aligned}
\tag{13.22}
$$

This contains only one dimensioned constant, qk/m, when there is no electric field. Since the SI system of units is based on four basic units (charge, mass, length, and time), we can absorb up to four constants by choosing different units. We choose units such that $qk/m=1=k$ and then equation 13.2 becomes

$$
\begin{aligned}
a_x &= v_z\imath \\
a_y &= -v_zy \\
a_z &= E-xv_x+yv_y \quad .
\end{aligned}
\tag{13.23}
$$

Similar simplifications occur in the difference equations 13.20 when the specifications of a quadrupole magnetic field and constant electric field E_z are introduced. These simplified equations have been used to produce the worksheet `EMQD.wk1` that is available in the student software package.

From equation 13.22, by comparing the terms $a_z=\ddot{z}$ and $(qk/m)v_yy$, we see that the constant $(qk/m)^{-1}$ has dimensions of length times time. It can be used therefore, to convert a velocity into $(\text{time})^{-2}$, i.e., a frequency squared. In this problem, then, there is a time scale $\sqrt{m/qkC}$ associated with any typical velocity C (or with any energy per unit mass $E/m=\frac{1}{2}v^2$). There is also a length scale $\sqrt{mC/qk}$ associated with a velocity C. In the `EMQD` worksheet, it is assumed that some typical velocity C has been selected as a reference and that the corresponding length and times scales are taken as basic units.

Exploration

By specifying different initial positions and velocities, you can have the work-sheet compute and graph different possible orbits for a charged particle in this nonuniform magnetic field. It is not necessary to try different magnitudes for the initial velocity, as this merely reproduces, on a different scale, a solution that can be found by scaling the initial position. You will find it necessary to use the {Alt-M} macro (see Appendix C at page 208) to do more integration than there is room for in the work block. If, however, the I/O screen reports a large energy error for the most recent integration, it is better to reduce the time-step dt and press {Calc} to repeat the trajectory you had just calculated at higher accuracy before continuing that trajectory with {Alt-M}.

Sample initial conditions that you might try first are sets $(x, y, z; v_x, v_y, v_z)$ with values $(2, 0, 0; 0, .6, .8)$, $(1, 0, 0; 0, .8, .6)$, and $(.1, 1, 0; 0, .8, .6)$. Note from equation 13.17 that $|\mathbf{B}| = k\sqrt{x^2 + y^2}$ increases away from the origin, so we can expect trajectory behavior in strong fields far from the origin to be different from that in weak fields near the origin. You may be able to find examples of magnetic containment, where charged particles appear to be trapped in a finite region by the magnetic fields, as is desired in efforts to produce nuclear fusion in the laboratory. You will probably also see examples of leakage, where particles escape by moving nearly parallel to the magnetic field lines. Trajectories that pass close to the origin tend to be chaotic, in the sense that there is no long-term pattern, since the charged particles are not held to the pattern of the field lines when the field is weak.

Part III

Appendices

Appendix A

Doing DOS

A.1 Major DOS commands

dir   Presents information about the files in the current directory.

copy   Copies specified files from one drive/directory to another.
del   Deletes specified files from a disk.
ren   Renames a file.
diskcopy Makes an exact copy of an entire floppy disk to another.

md   Makes (creates) a new directory with the name specified.
cd   Specifies which directory is to be used (or reports which is current).
rd   Removes (deletes) the specified empty directory.

format   Prepares a disk to receive data, deleting all previous data.
chkdsk   Checks and reports the space used and available on a disk.

date   Sets the internal clock to the date provided (e.g., 9-20-90).
time   Sets the internal clock to the specified time (e.g., 20:54).

type   Displays the contents of the specified text file.
             ({Ctrl NumLock} to pause; any key to resume; {Ctrl Break} to
             abort.)

A.2 Some disk filenames

`george.txt` A file named "george" whose extension is "txt". File names are
             limited to 8 characters, and extensions range from 0 to 3 characters.

`a:\ma*.*`   All files in the root directory of drive "a:" the names of which
             begin with "ma" and which have any extension.

`a:martha.exe`
> An executable program file named "martha" in the current directory on drive "a:".

`\sci\nuc\boron\info.doc`
> A file named "info" with extension "doc" in a subdirectory named "boron". "Boron" is itself a subdirectory of the subdirectory "nuc", which is in turn a subdirectory of "sci" which is, in turn, a subdirectory of the current disk drive's root directory.
>
> "\sci\nuc\boron\" is the "path" for this file. The path plus the file name and extension is called the "pathname" for the file.

Appendix B

Quick Guide to 1–2–3

B.1   Major 1–2–3 Commands

/Worksheet
Global Insert Delete Column Erase **Titles** Window Status Page

> **/Worksheet Global**
> **Format** Label-prefix Column-width **Recalculation Protection**
> Default Zero

/Range
Format Label **Erase Name** Justify Protect **Unprotect** Input
Value Transpose

/Copy

/Move

/File
Retrieve Save Combine Xtract Erase List Import **Directory**

/Print Printer
Range Line **Page Options** Clear **Align Go** Quit

> **/Print Printer Options Other**
> **As-displayed Cell-formulas** Formatted Unformatted

/Graph
Type **X A B** C D E F Reset **View** Save **Options Name** Quit

    /Graph Options
    Legend Format Titles Grid Scale **Color** B&W Data-labels Quit

    /Graph Name
    Use Create Delete Reset

/Data
Fill Table Sort Query Distribution **Matrix** Regression Parse

/System

/Quit

B.2   Some Important Mathematical Functions

@abs, sin, cos, tan, asin, acos, atan, exp, log, sqrt, pi, rand

@avg, count, max, min, sum, if

B.3   Function Keys in 1–2–3

F1	{Help}	about current menu selection, etc.
F3	{Name}	menu of files, ranges, or graphs
F5	{GoTo}	cell address or range name
F7	{Query}	(used in databases)
F9	{Calc}	recalc all (or cell while editing)

	F2	{Edit}	revise current cell
	F4	{Abs}	cycle cell's ab$olute address in formula
	F6	{Window}	jump to other window (when screen split)
	F8	{Table}	repeat last **/Data Table** command
	F10	{Graph}	show current graph

Note: {Name}, when pressed more than once, toggles between single line and full screen listing. {Abs}, when pressed more than once, changes a cell address in the following cycle: `C10`, `$C$10`, `C$10`, `$C10`, `C10`.

Appendix C

A Macro Zoo

A "macro" is a keystroke sequence that can be replayed easily. In Lotus 1–2–3 these keystrokes are written out in a few consecutive cells in a single column, using a notation almost identical to that we have adopted for prescriptions. Then, by giving the first cell the name **\x**, the playback is assigned to an {Alt-x} key where x is some alphabetic character. (A blank cell in the column marks the end of the macro.) The macros found most often in worksheets that accompany this book are:

{Alt-S} Saves the worksheet in compact form.

{Alt-X} eXpands the worksheet when compacted for saving.

{Alt-M} More integration, copies final state to initial state.

{Alt-R} Records data from a calculation on the notes page.

{Alt-W} creates two Windows, viewing I/O and workblock respectively.

{Alt-C} Clears windows back to standard screen.

Some of these are described in more detail below. If the computer beeps when you press one of these key combinations, it means that your worksheet does not include any macro with that key assignment, as many of our worksheets do not.

Using macros is quite easy until they do something unexpected, at which point it is very helpful to understand exactly what the macro *is* doing. You can inspect any macro in complete detail by going to its beginning cell (e.g. {GoTo}\S{Enter} for the {Alt-S} macro) and reading it. You can watch a macro execute by pressing {Step}, which is the {Alt-F2} key combination, before you execute the macro. It will then execute one step at a time, requiring you to tap the space bar for each successive step.

Writing macros is a more advanced form of programming than is writing worksheet formulas, and is only recommended to students who enjoy exploiting the computer more fully.

C.1   The \Save and \Xpand Macros

The \Save Macro

Most of the worksheets contain a long workblock that is almost entirely copies of a typical row near the top of the block. To save space on disk (and time in saving and retrieving the worksheet) it is desirable to delete most of these copied rows before saving the worksheet, and then reconstitute them whenever the worksheet is retrieved. The \S macro which is executed by pressing {Alt-S} does the saving part of this job. The most important part of using it is to realize that the filename under which the file will be saved is part of the macro. It appears in the cell named SAVENAME which you can find using {GoTo}, the F5 key. If you modify a distributed worksheet and wish to save it under a different name (to avoid overwriting the original), you must either use the standard /File Save command (no compacting), or else change the filename in the SAVENAME cell before using {Alt-S}.

The {Alt-S} macro always saves the file in the directory where it was most recently saved or retrieved. You cannot change this by writing a pathname in the SAVENAME cell. Instead you can first save the file in uncompacted form in the desired directory using /File Save, and then recover the excess disk space by saving a second time under the same name using {Alt-S}. If the disk doesn't have enough space for the temporary uncompacted file, save the file first to the previous directory with {Alt-S} (but Cancel the save if you are offered that option when the macro halts) which will compact the worksheet. Then save it using /File Save in the desired new directory. When you are modifying a worksheet, these annoyances only occur on the first save in a new directory. Thereafter you can continue to save improvements in that directory by simply pressing {Alt-S}.

After {Alt-S} has executed, the worksheet is in a compact form and will not operate as expected—most of the workblock is missing. To continue working, you must either retrieve it again or else execute the {Alt-X} macro to re-expand it.

The \Xpand Macro

The {Alt-X} and {Alt-S} macros are a matched pair. One is of little use without the other. When a file that has been saved with {Alt-S} is retrieved, the {Alt-X} macro executes automatically and re-expands the worksheet to its operational form. This occurs because {Alt-S} has given a second name \0, in addition to the permanent \X name, to the cell where the \Xpand macro begins. This \0 macro name has only one function—it identifies the macro as one that the spreadsheet should execute immediately upon retrieving the worksheet.

Since its operation in normally automatic, you need notice the {Alt-X} macro only under two conditions: you have saved a worksheet with {Alt-S} but wish to continue working on it, or you want to change the size of the point is constant, its acceleration is not zero. Rather its acceleration is directed radially inward and has a magnitude $\omega^2 R$. So equations 12.2 and 12.3 together describe a motion in which each point on the rope repeatedly maps out a circle of radius $R(h)$ with a common period $P = 2\pi/\omega$.[1] The minus sign makes the rope rotate clockwise as seen from above. This circularly polarized standing wave represents a very stable motion for the hanging rope and allows a simple analysis of the forces needed to maintain the standing wave. This analysis leads us to a differential equation for $R(h)$, which we then solve numerically.

12.2   A Time-Independent Wave Equation

We can obtain the differential equation for $R(h)$ by applying what we already know about uniform circular motion to a short segment of our rope. If the segment has a length Δh then its mass is $\mu \Delta h$ and its centripetal acceleration is $-\omega^2 R(h)$. To keep the segment moving in its circular path, Newton's second law tells us that there must be a net centripetal force

$$F = -\mu \Delta h\, \omega^2 R(h) \tag{12.4}$$

acting on it. This force is directed horizontally inward toward the equilibrium position of the rope and is the vector sum of gravity plus the forces exerted on each end of the segment by the rope immediately above and below it. The vertical components of these two forces can be taken as the tensions at the two points[2] and almost cancel; their small difference just balances the gravitational force on the segment

$$T_2 - T_1 = \Delta M\, g = \mu\, \Delta h\, g \quad . \tag{12.5}$$

The horizontal (radial) components

$$T_r = T(dR/dh) = \mu g h(dR/dh) \tag{12.6}$$

given by $+T_2(dR/dh)_2$ acting on the top of the segment and $-T_1(dR/dh)_1$ acting on its bottom, add to produce the required centripetal force

$$F = +T_2(dR/dh)_2 - T_1(dR/dh)_1 \quad . \tag{12.7}$$

Equating the right-hand sides of equations 12.4 and 12.7, then dividing by Δh and taking the limit as Δh approaches zero gives us

$$(d/dh)[T(dR/dh)] = -\mu\omega^2 R(h) \quad . \tag{12.8}$$

---

[1] If you are familiar with complex numbers you will be able to verify that equations 12.2 and 12.3 above can be combined into one: $x + iy = R(h)e^{-i\omega t}$.

[2] The moving rope makes a small angle φ with the vertical, so the vertical components of the tension forces at the ends are actually $\pm T \cos \varphi$. But since we assume a very small amplitude wave, φ will be small, and we approximate $\cos \varphi = 1$. Similarly in calculating the horizontal components of the tension forces we approximate $\sin \varphi = \varphi = \tan \varphi = dR/dh$.

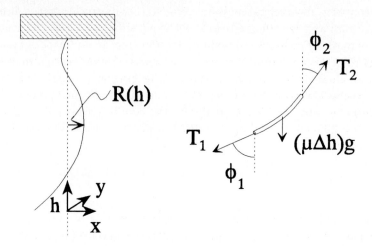

Figure 12.1: Hanging rope. The sketch on the left marks the horizontal displacement $R(h)$ of the moving rope at a height h above the bottom of the rope when $\cos \omega t = 1$. On the right is a magnified view of a small segment of the rope showing all the forces acting on this segment. They are the forces due to tension, which act on the ends, plus the gravitational force.

Wave Kinematics

Describing Standing Waves   The waves we are discussing are transverse waves; the motion of the rope is side to side as the wave propagates up or down the vertical rope. Since we are interested in the normal modes of the rope, we seek solutions in which the horizontal motion of each small segment of the rope is sinusoidal in time with some definite angular frequency ω. We also require that the vertical motion of each point on the rope be negligible, a condition that will be satisfied for small amplitude waves. Such motion of the rope is described by

$$x(h,t) = R(h) \cos \omega t \qquad (12.2)$$

which corresponds to the sketch in Figure 12.1.

Circular Polarization   Nothing restricts the rope to move only in the h-x plane. For example, the rope can simultaneously vibrate in the y-direction perpendicular to x and h, at the same frequency and with the same amplitude, satisfying

$$y(h,t) = -R(h) \sin \omega t \qquad . \qquad (12.3)$$

Why this particular choice of sines and cosines? The answer lies in the fact that $x = R \cos \omega t$ and $y = -R \sin \omega t$ describe the motion of a point along a circular trajectory of radius R at a constant speed ωR. Though the speed of the workblock. In the first instance just press {Alt-X} and the worksheet will re-expand. No further expansion occurs if you press {Alt-X} a second time.

Some precautions are necessary if you change the size of the workblock and want to continue using {Alt-S} and {Alt-X}. These macros rely upon markers in the first and last workblock columns for their operation, and will not work properly if these columns are deleted or have data /Moved on top of them. Also, columns you insert or add before the first workblock column, or after the last, will not be properly reconstituted by the {Alt-X} macro. Instead insert additional columns as needed somewhere between these outside columns. Adding *rows* with /Worksheet Insert Row is ineffectual since the next /File Retrieve will restore the original number of rows. Instead the \X macro must be modified as described below. Deleting rows near the top or bottom of the workblock gives unpredictable (i.e., possibly destructive) results. Graph ranges are usually attached to the top and bottom calculation rows, and {Alt-S} and {Alt-X} presume that STDLINE, a row near the top, is a typical row whose formulae will not be converted to @ERR when deletions occur a few rows below. /Moveing data onto a critical row, column, or cell is just a bad as deleting. Any range names attached to these locations become invalid. But changing formulas or /Copying to such cells or ranges causes no problems.

The best way to change the number of rows in a workblock that is to be managed by {Alt-S} and {Alt-X} is to change one number in the \X macro. Use {GoTo} to find the cell named \X where the macro begins. Edit that cell, whose contents should end with {PgDn 15}~ with 15 replaced by some other number. Change the number; each unit you add or subtract changes the size of the workblock by one "page" or screenful of 20 lines. The change takes effect only after you use {Alt-S} and {Alt-X} once. Modifying the worksheet this way changes its size in active memory, but does not change its size when stored on disk.

C.2   Other Macros

Other macros listed on page 205 are simpler to read than the \Save and \Xpand macros, and perform less complicated functions.

The \Window and \Clear Macros

The \Clear macro found, as a companion to \Window, in the worksheets from chapter 11 simply issues a single command /Worksheet Windows Clear followed by the keypress {Home}. This can all be typed into the single cell named \C as specifications for five keystrokes: '/wwc{Home}. When writing a macro you must type the label-prefix character "'" (or """ or "^") first so that the subsequent typing is entered as a label in the cell that is to contain the macro, rather than being executed as you type it. The \Window macro looks more complicated, but the essential part is just the two commands /Worksheet

Windows Horizontal /Worksheet Windows Unsynchronize written as eight keystrokes \wwh\wwu. The rest is mostly positioning the cell pointer (highlight) before and after issuing these commands. It is of course the bother of doing all that cursor movement to get the desired data together on screen that makes it worthwhile to have written the macro. This {Alt-W} macro automates much of the procedure described in exercise 7.11 on page 100, but is first used as a macro in the OSC worksheet from chapter 11.

The \More and \Record Macros

These macros use the **/Range Value** command to copy data from one place to another. The \More or {Alt-M} macro copies numbers computed in the last line of the workblock into the initial value cells on the I/O page so that a recalculation will continue the time evolution of a mechanical system. It begins (in cell \M) with **/rvX_LAST˜X_INIT˜** and continues with similar commands. (The tilde symbol ˜ here is Lotus's abbreviation for {Enter} in macros.) The difference between **/Range Value** and **/Copy** is that only the current numerical (or string) value of the cell is copied, not the formula by which it had been calculated. The target cell then does not change upon subsequent recalculation, which may change the cell from which the copy was made. Data that {Alt-M} does not change, such as the time step **dt**, can be modified before pressing {Alt-M}; this allows some control in situations where very different time steps are needed in different parts of an orbit. For instance, a comet orbit needs a small time step (days or weeks) when it is inside the Earth's orbit, but can use longer steps (years or decades) when it is beyond Neptune.

The \Record macro is used to build a table on the notes page. A row of formulae with the results it is desired to tabulate is written on the notes page and given the name **DATA**. The {Alt-R} macro then copies the values produced by these formulae to the first blank row in the table below. (It fails, however, if there is a blank row immediately below the **DATA** range.) This process was done without macros after the prescription on page 54 in chapter 3. As a macro it appears, for instance, in the OSC worksheet from chapter 11.

C.3   The Lotus Macro Language

In reading or writing macros, there are a few differences from the notation we have used in prescriptions in this text that must be noted. Although macros are essentially recorded sequences of keystrokes, only the character keys, the cursor keys, and the function keys can be recorded. The few keys that cannot be recorded are {CapsLock}, {NumLock}, {ScrollLock}, and {PrintScreen}. Macros are customarily written with the assumption that {ScrollLock} is toggled off before the macro is invoked. The function keys are only recognized when identified by name, e.g., {GoTo} is valid but {F5} will fail. The valid names are those in section B.3. The enter key cannot be written as {Enter}, but is represented by the tilde symbol "˜".

More complex macros can contain control statements, as in other computer languages. Our macros only use one such statement, an *if* construction. Macro \X begins '{if COMPRESSED˜}... (or /xiCOMPRESSED˜... as an alternative notation) which has the following effect: if the cell named COMPRESSED has a nonzero ("true") value, then the remaining commands or keystrokes (...) in the cell \X are executed, otherwise the macro skips on to the cell below. After successfully expanding the worksheet, the \X macro stores a zero ("false") in the cell named COMPRESSED; a second invocation of {Alt-X} then does not execute the *if*-controlled statements (...) which include the /wir (/Worksheet Insert Row) command that expands the size of the workblock.

C.4   Macro Listings

```
                              range names:
                              X_INIT  E11      V_INIT  G11      T_INIT  C11
                              X_LAST  M463     V_LAST  N463     T_LAST  L463
                              \M      I136
I135: U [W72] 'macro \M "More: continue integration to subsequent times"
I136: PR [W72] '/rvX_LAST˜X_INIT˜/rvV_LAST˜V_INIT˜/rvT_LAST˜T_INIT˜
I137: PR [W72] '{Calc}{Graph}


C20: U ˜macro \R "record data"                      range names:
C21: PR '{GoTo}DATA˜{End}{Down}{Down}               \R    C21
C22: PR '/rvDATA˜˜                                   W     D7
C23: PR '{GoTo}A5˜{GoTo}W˜                           X_MAX D19
                                                    DATA  C26..G26
C19: PR "X_max =        D19: PR @MAX(XALL)
C25: PR ˜data records:           E25: U 'name & date      Q=10 G=1.00
C26: (T) PR +W    D26: (T) PR +X_MAX   E26: (T) PR +X_INIT
                  F26: (T) PR +V_INIT  G26: (T) PR -@ATAN2(X_INIT,V_INIT/$W)
C27..G27: (T) PR \-
```

Listing C.1: The \More and \Record macros as they appear in the OSC worksheet.

```
I126: U  [W72] 'macro \W "make two Windows"         range names:
                                                    \W        I127
I127: PR [W72] '{End}{Home}{GoTo}A5˜                T         L143
I128: PR [W72] '{Down 12}/wwh/wwu                   \C        I133
I129: PR [W72] '{Window}{GoTo}T˜{Up 2}
I130: PR [W72] '{Down 3}/wth{End}{Down}{Up}{Window}{GoTo}W˜

I132: U  [W72] 'macro \C  "Clear to single window"
I133: PR [W72] '/wwc{Home}
```

Listing C.2: The \Window and \Clear macros as they appear in the OSC worksheet.

```
I96:  U  [W72]  "COMPRESSED                                    range names:
I97:  PR [W72]  0                                             COMPRESSED  I97
                                                              \X          I102
I98:  U  [W72]  'The following Macro, named  \0 and \X  is    \0          I102
I99:  U  [W72]  'executed automatically each time
I100: U  [W72]  'this worksheet is retrieved.
I101: PR [W72]  'Macro \0 or \X

I102: PR [W72]  '/xiCOMPRESSED~{GoTo}STDLINE~{Down 2}/wir{Down 14}{PgDn 15}~
I103: PR [W72]  '{GoTo}COMPRESSED~
I104: PR [W72]  '/cSTDLINE~COPYTARG~                          range names:
I105: PR [W72]  '/cZERO~COMPRESSED~/rnd\0~                    STDLINE    L145..T145
I106: PR [W72]  '/ruLASTLINE~{Home}                          COPYTARG   L146..L463
I107: PR [W72]  '{GoTo}W~/wgpe{Calc}                         LASTLINE   L463..T463

I109: U  [W72]  'A subroutine in Macro \S compresses this
I110: U  [W72]  'worksheet before it is saved.

I112: U  [W72]  "TRUE                                         range names:
I113: PR [W72]  1                                            TRUE       I113
I114: U  [W72]  "ZERO                                        ZERO       I115
I115: PR [W72]  0


I117: U  [W72]  'macro \S "Save this worksheet"

I118: PR [W72]  '/wgpd
I119: PR [W72]  '/xi#not#COMPRESSED~{GoTo}STDLINE~{Down 2}/wdr{End}{Down}{Up 2}~
I120: PR [W72]  '/cTRUE~COMPRESSED~/rnc\0~\X~
I121: PR [W72]  '{Home}{GoTo}TITLE1~                          range names:
I122: PR [W72]  '/fs                                         \S         I118
I123: U  [W72]  'OSC                                         SAVENAME   I123
I124: PR [W72]  '~
```

Listing C.3: The \Save and \Xpand macros as they appear in the OSC worksheet.

Appendix D

Optional Topics in Spreadsheet Physics

D.1   MOTION: Kinematics in One Dimension

In equation 2.3 and nearby we discussed computing derivatives numerically by using finite differences, but the important applications that followed used these formulae for integration—predicting change from known rates. Here we return to the descriptive side of these equations: assume the motion $x(t)$ is known and use numerical methods to describe its derivatives v and a. Schema D.1 shows how this would be arranged in a worksheet, and Listing D.4 at the end of the chapter gives details for constructing a worksheet MOTION that will calculate and plot the velocity and acceleration (calculated this way) from any function $x(t)$ that you enter into it. (This worksheet is also found in the student software package that is available through your instructor.) Experimental data could be entered into the $x(t)$ column if they are available at regular time intervals dt, or as t, x data pairs otherwise. The exercises below illustrate how analytic formulae can be used in the MOTION worksheet, even if you haven't yet learned enough calculus to form the derivatives analytically.

EXERCISE D.1^{E} In the MOTION worksheet, install $x(t) = x_0 + v_0 t - (1/2)gt^2$ in the x column. Note that the worksheet has positions available for storing x_0, v_0, and g. The essential steps are given by the following prescription:

```
D18: +$C$7+$C$8*T-$C$9*T^2/2
D18: /Copy {Enter} {Down}{Period}{End}{Down}{Enter}
 --: {Calc}{Graph}
     Install the x(t) formula you want, recalculate, and look at a graph. The
     {Calc} and {Graph} keys are {F9} and {F10}.
```

Print a plot (the graph is available under the name XVA_T) showing x, v and a for initial conditions x_0 and v_0 that show all three clearly on one plot. Explain how it agrees with analytic formulas for v and a that you can derive.

labels:	t	x	vhs	a	v
init data:	t_0: 0	x_0: $x(t_0)$	$v_{1/2}$: $(x_1 - x_0)/dt$		
typical line:	t_1: $t_0 + dt$	x_1: $x(t_1)$	$v_{3/2}$: $(x_2 - x_1)/dt$	a_1: $(v_{3/2} - v_{1/2})/dt$	v_1: $(v_{3/2} + v_{1/2})/2$

Schema D.1: Kinematics in one dimension. For any position description $x(t)$ tabulated in the **x** column of this table, the next two columns calculate the velocity and acceleration. Velocities at times "in step" with the position and acceleration tables can be computed as in the last column by averaging the "half-step" velocities in the **vhs** column. (This is needed only if the time steps dt are large enough to show a distinction between these two velocity approximations.)

EXERCISE D.2^E In the MOTION worksheet, install $x(t) = A\cos(\omega t + \phi)$ in the **x** column. Let A, ω and ϕ be parameters whose effect you can study by varying them and comparing plots. Note in the XVA_T graph that analysis is much simpler if you take $1/\omega$ as the unit of time (i.e. set $\omega = 1$ in the computations). Print such a graph after entering useful title information in the first two rows of your worksheet, and pencil on it the values of the parameters you used. Then choose a large value of dt such as $\omega\, dt = \pi/6$ and compare (print out) the V_X plot for the case, as given, where it plots velocity from the **v** column, and for the case (change the graph range plotted) where it plots velocity from the **vhs** column which is a half step $(dt/2)$ out of step with the positions in column **x**.

EXERCISE D.3^E In worksheet MOTION install a sample motion $x(t)$ using the formula

$$x(t) = uT\,[(1 - t/T)\ln(1 - t/T) + t/T] \tag{D.1}$$

which is a theoretical result for the motion of a rocket launched outside the Earth's gravitational field. The velocity constant u represents the exhaust velocity of expelled gas relative to the rocket. The time T is chosen so that M_0/T is the (constant) rate in kg/s at which fuel is burned if M_0 is the initial mass of the fueled rocket. (Thus $M(t) = M_0(1 - t/T)$ is the mass of the rocket at any time, and clearly only $t < T$ makes sense since the rocket must have some parts that are not fuel.)

Use (at first) values $u = 2800\,\text{m/s}$ and $T = 400\,\text{s}$ and vary dt to see how the limit $dt \to 0$ works in numerical differentiation. Nonsense at very small dt is due to roundoff error, i.e. to the fact that your computer only carries about 14 significant figures when it calculates. At what largest size dt can you document (print a graph) the first clear signs of this problem? For some much smaller dt such as $dt = 10^{-30}\,\text{s}$ explain how the computer arrived at the value it gives for the first velocity in the table.

Add a column to the worksheet to calculate the quantity $M(t)\,a/M_0$ to check that the rocket's thrust (the force causing the acceleration via $F = Ma$) is constant.

D.2 GTPLT: **Function graphing template**

There will be many occasions when you will meet unfamiliar analytic functions that are used to describe force laws or potential energies. In most cases you will understand a function's properties better and more easily by graphing it before you try to use it in physics applications. The force laws suggested in equations 5.5 and 5.6 are examples. (Their physical significance is not considered here; see Chapter 5.) In this section we will graph these algebraic functions, and in the process produce a worksheet **GTPLT** that you can file away as a graph template to be modified any time you need a quick look at the properties (maxima, minima, zeros, singularities, asymptotic behavior) of a newly encountered function.

Once it can be done in just a few minutes, making a graph becomes an efficient way to begin exploring the properties of an unfamiliar function. You can use this technique not just in physics courses, but in calculus, engineering, chemistry, economics, or wherever you meet an unfamiliar mathematical expression. When you need a graph, edit the saved template by typing in the new function, explore for the value ranges that are significant, and print out a useful version.

Tabulating the function Recall from Chapter 1 that the procedure to graph a function $f(x)$ is first to make a table with the independent variable x in one column, and the dependent variable f in a parallel column, and then to specify these columns for graphing in the /**Graph** menu. For definiteness let us first graph the function

$$F_3 = \frac{1}{2} \left(\frac{1}{x^5} - \frac{1}{x^3} \right) \tag{D.2}$$

from equation 5.5. You might be tempted to make your worksheet as simple as Screen D.1 whose cell contents are listed beside it. That would not save time, however, as you would likely not have needed the graph if you already knew the most interesting range of x values to plot. Therefore plan to explore various

C1: (B1^-5-B1^-3)/2

	A	B	C
1		0.1	49500
2		0.2	1500
3		0.3	187.2427

B1: 0.1   C1: (B1^-5-B1^-3)/2
B2: +B1+0.1 C2: (B2^-5-B2^-3)/2
B3: +B2+0.1 C3: (B3^-5-B3^-3)/2
B..: ...   C..: ...
B51: +B50+0.1 C51: (B51^-5-B51^-3)/2

Screen D.1: Oversimplified (impractical) tabulation, with Listing D.1.

possibilities by using our standard format: title page, I/O page, calculation. (For scratch work that you will not even save for yourself, the title page might be omitted, but not the I/O page.) A worksheet in this format is shown in

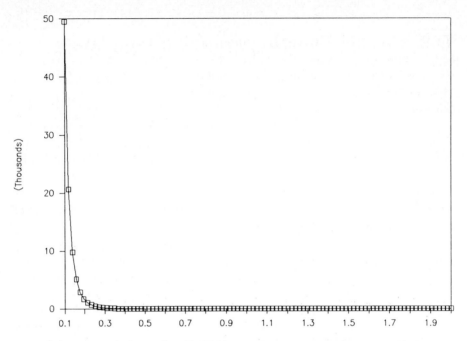

Figure D.1: The function F_3 when plotted between the limits specified in Screen D.2.

Screen D.2, where the "pages" in this simple case are just short areas a few lines long. The I/O page is just two rows **6** and **7** providing a place to enter the minimum and maximum values of x that you want shown on your graph. There is also included nearby a computation of the step size dx that will be used in the tabulation of the function to provide data for plotting. Apart from labels the construction is:

```
F7:  (C7-C6)/100
F7:  /Range Name Create dx{Enter} {Enter}
B10: +C6
B10: /Range Name Create x{Enter} {Enter}
C10: (x^-5-x^-3)/2
C10: /Copy {Enter} {Down} {Enter}
B11: +x+$dx
B11: /Copy B11..C11 to B12..B110
```

After completing this tabulation (Exercise D.4), you should proceed to draw the graph (Exercise D.5) as described below.

Graphing   The next chore is to make a graph of the function you have just tabulated:

```
B10: /Graph Type XY
```

```
C10: (X^-5-X^-3)/2                                                    EDIT
(B10^-5-B10^-3)/2
```

	A	B	C	D	E	F	G	H
1	your name & today's date							
2	your class & section			Graph of F_3				
3								
4				F_3 = (1/x^5 - 1/x^3)/2				
5	--							
6		xmin =	0.1					
7		xmax =	2		dx =	0.019		
8	--							
9		x	F_3					
10		0.1	49500					
11		0.119	20655.76					
12		0.138	9799.955					
13		0.157	5112.492					
14		0.176	2869.070					
15		0.195	1705.928					
16		0.214	1063.022					
17		0.233	688.5711					
18		0.252	460.7583					
19		0.271	316.9542					
20		0.29	223.2688					

```
GRF3.WK1
```

Screen D.2: A tabulation ready for graphing.

> /G: X B10..B110
> /G: A C10..C110
> /G: Options Color Quit
> /G: View Quit
> This should be straightforward. Remember to use {End} {Down} to point out the tabulated ranges simply in the /Graph X and /Graph A commands.

The point of making this graph, of course, is to understand the behavior of the F_3 function, very little of which is found in the first graph you look at (Figure D.1). You must now vary the limits x_{min} and x_{max} to explore regions that are not overwhelmed by the large function values at small x to see how the function behaves elsewhere. (Using manual scaling via the menu /Graph Options Scale Yscale is also effective, but probably more work.)

A reusable graph template

The graph you have made thus far was done quickly, is serviceable, but somewhat unfinished. If we polish it up now, with a thought to using it again later for other functions, then any improvements now will be available at no cost in later applications. Let us then name a few cells for convenience, and put titles on the graph that will reflect the labels in th   ksheet which could be easily

```
F7: (XMAX-XMIN)/100                                                    EDIT
(C7-C6)/100
```

	A	B	C	D	E	F	G	H
1	your name	&	today's	date				
2	your class	&	section		Graph of F_4			
3								
4					F_4 = (x/2) (1 - x^2) [5 / (2 + 3x^2)]^3			
5	------	------	------	------	------	------	------	------
6		xmin =	0.99					
7		xmax =	1.01		dx =	0.0002		
8	------	------	------	------	------	------	------	------
9		x	F_4					
10		0.99	0.010211					
11		0.9902	0.010003					
12		0.9904	0.009795					

Screen D.3: GTPLT, a template for simple graphs.

changed for different functions.

```
     --: {Home}
     A1: /Range Name Create title2{Enter} {Enter}
     D2: /Range Name Create title1{Enter} {Enter}
     C6: /Range Name Create xmin{Enter} {Enter}
     C7: /Range Name Create xmax{Enter} {Enter}
     B9: /Range Name Create arg{Enter} {Enter}
     C9: /Range Name Create value{Enter} {Enter}
```
Assign a batch of cell names.

```
     --: /Graph Options
    /GO: Titles First \title1{Enter}
    /GO: Titles Second \title2{Enter}
    /GO: Titles X-axis \arg{Enter}
    /GO: Titles Y-axis \value{Enter}
```
Assign graph main titles from the title page of the worksheet, and axis titles from the column headings.

```
    /GO: Format A Lines Quit Quit View
     /G: Quit
```
Remove the data point markers for a smoother appearance.

If you save the worksheet at this point, call it GRF3.

Let us now see how easily the worksheet can be changed to plot a different function. Change the title page (rows **1** through **4**) to give a current date and the heading information for the F_4 function from equation 5.6 as in Screen D.3. Also modify the input data, x_{min} and x_{max}, as suggested there. Then rebuild the tabulation:

```
     C9: F_4
```

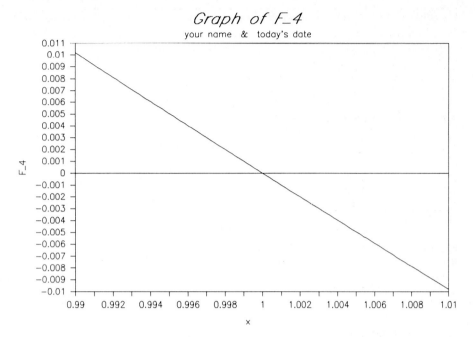

Figure D.2: Any smooth function looks like a straight line when graphed over a small interval, as in this plot of F_4 from the GTPLT worksheet.

```
C10: (X/2)*(1-X^2)*(5/(2+3*X^2))^3
C10: /Copy {Enter} {End} {Down} {Enter}
  --: {Graph}
```
       Recall that the {Graph} key is {F10}.

You should now see a graph like Figure D.2 with the appropriately changed titles inherited from the changes you made in the body of the worksheet. Again the x-range gives only a little information about F_4, and you must explore a variety of different choices before you get a good picture of this function's general behavior (Exercise D.6). Save this worksheet

```
  --: /File Save GTPLT
```

and reuse it frequently whenever you meet an unfamiliar function.

EXERCISE D.4^C Make the worksheet to tabulate F_3 as shown in Screen D.2 and Listing D.2.

EXERCISE D.5^C Add the graph settings to the worksheet you made in Exercise D.4 to produce the corresponding graph. As a check, the graph should look like Figure D.1 when you use the x limits shown in Screen D.2. To learn more about the function, try other limits. When you get a graph that shows behavior other than divergence for small x, print it.

EXERCISE D.6^C Build the GTPLT worksheet as described above, with F_4 installed as the function to be plotted. Check it against Figure D.2. Use {PrintScreen} and print better plots. Where are the zeros of F_4? Is its slope dF_4/dx positive or negative at these zeros? Is F_4 a bounded function? Is it symmetric under the exchange $x \rightarrow -x$?

D.3   Listings

```
A5..H5: \-
B6: "xmin =       C6: U 0.1
B7: "xmax =       C7: U 2
  E7: "dx =         F7: (C7-C6)/100
A8..H8: \-

B9: "x            C9: "F_3
B10: +C6          C10: (X^-5-X^-3)/2
B11: +X+$DX       C11: (B11^-5-B11^-3)/2
B..:   ...        C..:   ...
B110: +B109+$DX   C110: (B110^-5-B110^-3)/2
```

<pre>
                                          Graph Settings
                                 Name: F_3   Type: XY
                                 Range   Format
                                   X: B10..B110
                                   A: C10..C110   Both
</pre>

Listing D.2: In these formulae to tabulate F_3 note that the cell B10 has been assigned the name X, and that cell F7 has been named DX.

<pre>
                          Graph Settings
              Name: F_X   Type: XY
              Titles:
                1st: \title1
                2nd: \title2   Range   Format
              X-axis: \arg   X: B10..B110
              Y-axis: \value   A: C10..C110   Lines
</pre>

```
range names:              E7: "dx =      F7: (XMAX-XMIN)/100
TITLE2 A1    TITLE1 D2    B9: "x         C9: "F_4
XMIN   C6                 B10: +XMIN     C10: (X/2)*(1-X^2)*(5/(2+3*X^2))^3
XMAX   C7    DX     F7    B11: +X+$DX
ARG    B9    VALUE  C9                   C11: (B11/2)*(1-B11^2)*(5/(2+3*B11^2))^3
X      B10               B12..C110:   ...
```

Listing D.3: GTPLT, a template for simple graphs.

```
A1: 'your name  &  today's date
A2: 'your class & section                    Q1: 'Notes page:
       D2: 'MOTION: Kinematics in one dimension.    Q3: U 'range names:
           E3: 'Numerical differentiation              A        F19
           E4: 'by finite differences.                 DT       C13
A5: \-   ...   H5: \-                                   G        C9
A6: U 'Input parameters:                                T        C18
       B7: "x_init = C7: U 0        D7: 's             TITLE1   D2
       B8: "v_init = C8: U 20       D8: 'm/s           TITLE2   A1
       B9: "g =      C9: U 9.8      D9: 'm/s^2          V        G19
       B10: "w =     C10: U 1       D10: '/s            VHS      E18
A12: U 'Data rate:                                     V_INIT   C8
       B13: "dt =    C13: U 0.05    D13: 's             W        C10
           E13: 'Time step in the computation.         X        D18
A14: \-   ...   H14: \-                                 X_INIT   C7

A15: 'name:      C15: ^time    D15: 'distance
A16: 'units:     C16: ^[s]     D16: ^[m]
A17: 'labels:    C17: ^t       D17: ^x
A18: 'init data: C18: 0        D18: U +$X_INIT+$V_INIT*T-$G*T^2/2
A19: 'typical row: C19: +T+$DT  D19: U +$X_INIT+$V_INIT*C19-$G*C19^2/2
A20: 'copied rows: C20: +C19+$DT D20: U +$X_INIT+$V_INIT*C20-$G*C20^2/2
                 C..: ...      D..: ...
                 C117: +C116+$DT D117: U +$X_INIT+$V_INIT*C117-$G*C117^2/2
                 C118: +C117+$DT D118: U +$X_INIT+$V_INIT*C118-$G*C118^2/2

E15: ^velocity        F15: ^accel           G15: ^velocity
E16: ^[m/s]           F16: ^[m/s^2]          G16: ^[m/s]
E17: ^vhs             F17: ^a                G17: ^v
E18: (D19-X)/(C19-T)
E19: (D20-D19)/(C20-C19)    F19: (E19-VHS)/(C20-T)*2    G19: (VHS+E19)/2
E20: (D21-D20)/(C21-C20)    F20: (E20-E19)/(C21-C19)*2  G20: (E19+E20)/2
E..: ...                    F..: ...                    G..: ...
E117: (D118-D117)/...       F117: (E117-E116)/...       G117: (E116+E117)/2
```

Listing D.4: The MOTION worksheet. This worksheet calculates the kinematic variables in one dimension, namely x, v, and a, from any formula supplied in column D for the position $x(t)$. It implements Schema D.1, except that the dt denominators in approximating derivatives are computed in each case (to allow the time column to be replaced by observational data at unequal intervals) instead of being assumed constant.

Appendix E

Software Installation

E.1   PrintScreen Activation

The MS-DOS operating system supplies a utility program GRAPHICS.COM that makes {PrintScreen} work in graphics mode for a very limited combination of CRT displays and graphics printers. If this utility works with your hardware you should use it as explained in the DOS manual. In most cases, i.e., with EGA or VGA displays or with laser printers, it will be necessary to install a third party program that understands your hardware before you can print an image of a graphics screen. A program that works with a wide variety of displays and printers is GRAFPLUS. It is available from Jewell Technologies Inc. (4740 44th Avenue S.W., Suite 203, Seattle WA 98116; telephone 206-937-1081).

If you are unable to use {PrintScreen} for graphs, you can use the program PGRAPH that Lotus supplies with its 1–2–3 program. This program is explained in the spreadsheet reference manual. To use it you must first, within 1–2–3, use the command `/Graph Save` to save the graph you want as a separate `*.PIC` file on disk. In a second step, after you leave the 1–2–3 program, you run the PGRAPH program to translate the saved `*.PIC` files to your printer. Other spreadsheets may have built-in graph printing facilities in menu selections beyond those found in 1–2–3.

E.2   Student Software Package

Obtaining the Software Disk

We have prepared a collection of Lotus 1–2–3 worksheets to accompany this book. How you obtain the worksheets depends on how you purchased the book. If you are a student using this book in a course, your instructor will make the worksheets available to you in accordance with the course syllabus in return for the proof-of-license form from your copy of the book. Keep that form with the book until it is collected by your course instructor. (The disk will be provided to the instructor by Addison-Wesley upon adoption of *Spreadsheet Physics*.) If you have purchased *Spreadsheet Physics* as part of *The Student Edition of 1–2–3 for Physicists*, the disk of worksheets is included with the other disks in your package. Others users of *Spreadsheet Physics* who wish to purchase the *Spreadsheet Physics Worksheets* disk can order this Addison-Wesley product where they purchased the book.

Using the Software Disk

The distributed software disk should be write protected (by covering the notch on the 5.25 inch diskette with an opaque tab or by sliding the plastic tab on a 3.5 inch diskette until the write protect hole is fully open). The original disk should only be used to produce working copies of the worksheets on your hard disk or on other floppy disks and should be stored safely as backup in case the working copies become damaged.

The software is provided in compressed form in order to allow its distribution on a single diskette. The worksheets are packed into three self-extracting files named **student.exe**, **i1.exe**, and **i2.exe**. The file **student.exe** contains the student versions of the worksheets mentioned throughout the book. These sheets are usually only partially completed; you complete them following our instructions while you are learning the physics contained in each sheet. The illustrative sheets within **i1.exe** and **i2.exe** are fully developed examples from *Spreadsheet Physics*, often going well beyond what we discuss in the text. If you are learning this material for the first time, we recommend that you complete the student sheets before you turn to the illustrative sheets.

To unpack the compressed files, you must first prepare a destination for the worksheets. Each of the three compressed files produces up to 330 kilobytes of worksheet files, so your destination directory must have at least that much space available. There are many alternatives. The destination might be a subdirectory on your hard disk. It might be a floppy disk in a second floppy-disk drive. It could even be the same disk that holds your compressed files, provided you are using a 3.5 inch disk and are only unpacking one of the three compressed files. We recommend that you unpack the illustrative worksheets and the student worksheets into separate destination directories or disks.

Once you have established the destination, the process of unpacking a compressed file is easy. The general command is d:**student** *tpath* where d: is the drive containing the distribution disk and *tpath* is the pathname of the target directory. Replace **student** by **i1** or **i2** and use a different destination when unpacking the illustration packages.

For example, if you have two floppy drives, put the distribution disk in A: and place a newly formatted blank disk in B:. This blank disk may be any size that drive B: can use. Unpack the student software package by typing **a:student b:** to put the worksheets in the root directory of drive B:.

As another example, if your compressed disk is in drive A: and you want

to unpack the student worksheets into a subdirectory named `\SSPHYS` on your
C: hard drive, you should type, at the `C:>` prompt, `md \ssphys` to create the
destination directory and then `a:student c:\ssphys` to unpack the student
sheets. (For the files from `i1.exe` and `i2.exe` we suggest `C:\SSPHYS\I` as a
destination directory. It can be created by the command `md \ssphys\i` if you
have previously created the subdirectory `\SSPHYS`.)

As a third example, for a laptop computer with only a single 3.5 inch floppy
drive, make a diskcopy of the distribution disk, then use that copy and simply
type `i1` at the `A:>` prompt. This unpacks the first set of illustrative worksheets
directly onto the same disk that holds the three compressed files. The `*.exe`
files can be deleted from this working (copied) disk after the `*.wk1` files have
been produced by this unpacking process. You can repeat this entire process
for `i2.exe` and then for `student.exe` on separate copies of the distribution
disk to make three working floppy disks.

Software License

The original purchaser of this book is granted a nontransferable license to
copy the associated *Spreadsheet Physics Worksheets* for personal use only. The
licensee may not provide copies of these worksheets for use by anyone else.
Worksheets that incorporate substantial modifications by the licensee may,
however, be copied for use by others (e.g., by a course instructor or grader).
If you are a student using this book in a course, your instructor will make the
worksheets available to you in accordance with the course syllabus in return
for the proof-of-license form from your copy of the book. Keep that form with
the book until the form is collected by your course instructor.

Index

Proof of License
for
Spreadsheet Physics Worksheets

This form is a non-transferable license to copy for personal use the **Spreadsheet Physics Worksheets** associated with this book.

Students enrolled in courses using this book should submit this form to their instructor when requested.

Name_____

Student Number_____

Course_____

Section_____

Signature_____